LIGHT OF THE STARS

OTHER BOOKS BY ADAM FRANK

Astronomy:
At Play in the Cosmos

The Constant Fire:
Beyond the Science vs. Religion Debate

About Time:
Cosmology and Culture at the Twilight of the Big Bang

LIGHT
OF THE
STARS

ALIEN WORLDS AND
THE FATE OF THE EARTH

Adam Frank

W. W. NORTON & COMPANY
INDEPENDENT PUBLISHERS SINCE 1923
NEW YORK   LONDON

For information about permission to reproduce selections from this book, write to Permissions, W. W. Norton & Company, Inc., 500 Fifth Avenue, New York, NY 10110

For information about special discounts for bulk purchases, please contact W. W. Norton Special Sales at specialsales@wwnorton.com or 800-233-4830

Manufacturing by LSC Communications Harrisonburg
Book design by Daniel Lagin
Production manager: Lauren Abbate

Library of Congress Cataloging-in-Publication Data

Names: Frank, Adam, 1962– author.
Title: Light of the stars : alien worlds and the fate of the Earth / Adam Frank.
Other titles: Alien worlds and the fate of the Earth
Description: New York : W.W. Norton & Company, Inc., [2018] | Includes
    bibliographical references and index.
Identifiers: LCCN 2017061640 | ISBN 9780393609011 (hardcover)
Subjects: LCSH: Cosmology—Popular works. | Earth (Planet)—Popular
    works. | Human ecology—Popular works. | Exobiology—Popular works.
Classification: LCC QB982 .F73 2018 | DDC 523.1—dc23
LC record available at https://lccn.loc.gov/2017061640

W. W. Norton & Company, Inc., 500 Fifth Avenue, New York, N.Y. 10110
www.wwnorton.com

W. W. Norton & Company Ltd., 15 Carlisle Street, London W1D 3BS

1 2 3 4 5 6 7 8 9 0

To my sister, Elisabeth Frank, and our long, strange road.
I am grateful that your humor, resolve, and
Camp Dawson friendship were on this path with me.

CONTENTS

LIGHT OF THE STARS

INTRODUCTION

THE PROJECT AND THE PLANET

THE COSMIC TEENAGER

Imagine a room full of teenagers. The chairs are arranged in a loose circle, and the air smells of cheap cleaning products and anxiety. The kids are mostly in their late teens. Some are slumped in their chairs, trying to look bored; others lean forward, listening closely. They are here to tell their stories. The sixteen-year-old girl in the Black Sabbath T-shirt and chipped black nail polish got busted for dealing drugs at her high school. The skinny boy with a bad tattoo on his hand was arrested for joyriding in his grandparents' car. They're all in this room because they're on the wrong road. Old enough to have some power over their own lives, they've been making bad choices, destructive choices.

Each of the kids takes a turn unspooling how they got here. Some came from families that could barely hold it together. Others were trapped in their own feelings of isolation and insecurity. But in telling their stories, some of the kids catch a glimpse of an insight. It's something they couldn't imagine, couldn't truly *feel*, before.

They are not alone. They are not the first.

The circle and the stories give some of the kids the chance to see that it's not just them. Their individual stories are not so individual. Other kids their age have walked this road before, and some have even found a way out. Some have found a way to grow up.

• • •

WE HUMANS, with our project of civilization, are like those kids.

The massive collective project we call civilization began about almost ten thousand years ago, when the last ice age ended and our planet's climate grew warmer and wetter. In response, some of us stopped our nomadic wandering and settled into villages. Around those small groupings of huts and storehouses, the Earth was put to plow. We cultivated grains and rice. We domesticated the ox, the goat, and the cow. We created a new way of being human beyond the old hunter-gatherer way of life. It was an agricultural revolution that brought with it a radically different way of understanding ourselves and our place beneath the stars. This project of civilization accelerated when some of the villages grew into the first cities. There, we developed sophisticated new technologies for irrigation. We forged metals and stored information in writing. Through the tumult of markets and trade and conflict, our work became specialized. Some of us became millers, others tanners, others soldiers, and others still administrators. Some of us even became a special kind of priest whose job it was to watch the skies. And all the while, our numbers steadily increased. By one thousand years after the birth of Christ (1000 CE), three hundred million human beings walked the Earth.[1] Then, just five or six centuries ago, a new approach to the natural world was established. Harvesting ideas from across the planet, we discovered a method for directly probing the world's behavior— what we now call science. Using it, our project's capacities exploded. We learned to cross oceans quickly in ever-larger ships and in relative safely. Improvements in sanitation and medicines began to keep us from dying young. Machines for farming began freeing us from famine. In response, population growth rates exploded, and in the first half of the 1800s, our numbers crossed the one billion mark.[2]

In the years around that milestone, we made perhaps the most important discovery for our project of civilization. Using the fruits of our newly established scientific society, we learned how to harvest fossil fuels. Tapping a hundred million years of stored sunlight in the form of coal, and then petroleum,[3] industrial civilization tidal-waved across the globe. Touching even the most remote corners of every continent and every ocean, our capacities seemed to grow without limit. By 2011 CE, just about two centuries after reaching one billion, our numbers had climbed to seven billion.[4] Today, even a modest-sized modern city houses more people than lived on the entire planet before the dawn of agriculture. Using the tools of science and its daughter technologies, we explored the entire planet. We mapped everywhere. We *were* everywhere. These days, at any given moment, there are even half a million people flying miles above the ground.[5]

Our project was thriving.

For the most part, the planet took little notice of our experiment in civilization building. The clearing of land for farming certainly altered local balances of life and resources, but the Earth as a whole—meaning the surface, air, water, and life—wasn't significantly and globally disturbed from the state we found it in when civilization began. With the industrial revolution in the 1800s, the relationship between the project and the planet shifted. Earth began to "feel" our presence. Air, water, ice, rock—all the interdependent, strongly linked parts of the planet we inhabit—began to change. And, as it has done many times before in its four-and-a-half-billion-year history, the Earth started shifting from one planetary "state" to another.

The relatively temperate planet our project of civilization was born into began sliding into the past. Something new, something as yet unknown, now waits for its own time to begin. The planet is changing, and it's changing because of us. Those changes will, without doubt, stress our project of civilization. If the changes are extreme enough, they may even make the kind of civilization we rely on for survival impossible to continue. Our project may collapse.

And that is why humans, with our project of civilization, are cosmic teenagers (as Carl Sagan often noted). Our technology and the vast energies it has unleashed give us enormous power over ourselves and the world around us. It's like we've been given the keys to the planet. Now we're ready to drive it off a cliff. Unlike those kids, however, we are still blind to the truth. We are still unable to see the reality our project of civilization only recently revealed.

We are not alone. We are not the first.

Across our history, we have never seen our project of civilization—or ourselves, for that matter—as anything but a one-time story. We have always appeared to ourselves as something essentially new, something entirely different. Every step we humans took was a step into the unknown. There was nothing to guide us, no other histories we could look to and know what might be expected.

It is time to put that story to rest, because we have grown past it.

Through long effort, we have mapped out the four-billion-year history of life on Earth, and it shows us that we are not the first. We are not the first time a species has changed a planet's climate through its own success. The Earth and its inhabitants have been evolving together for eons, and we are just the most recent in a long line of its experiments.

But there is more.

Our science has also shown us something we did not know even twenty years ago. The universe is awash in planets and they are, in principle, not so different from our own. There is every reason to expect that on many of these worlds there will be oceans and currents. There will be mountains with fierce winds and valleys that begin the day shrouded in morning fog and end it with falling rain.

And there will be life, too. Sure, it is *possible* we're on the only world to host life in all of cosmic history. Science has, of course, been arguing about the existence of life on other worlds for centuries. But the explosion in our knowledge about other worlds sheds new light on this question, revealing something remarkable. The discovery of all those new planets means we can only be unique if the laws of the universe are strongly biased against life and intelligence. In other words, there are so many planets in the right place for life to form that the burden now falls on the pessimists. It's up to the naysayers to demonstrate how, with so many worlds and so many possibilities over the whole of cosmic space and time, we somehow are the first and the only.

So, while it is important to remember that the question of other life on other worlds remains open and undecided, we can now see that there has, most likely, been life before us. And on some worlds that life will, most likely, build rich and complex biospheres. Going further still, over the long history of the cosmos, life on some of those other worlds will, most likely, have woken up. It will have learned to think, to reason, and even to build its own projects of civilization.

One way or another, science points to the fact that we are likely not the first. Now it's time to take those insights from astronomy and earth science seriously. In light of our maturing knowledge, it's time to tell a different story about ourselves and our fate among the stars and their many worlds.

SCIENCE AND MYTH

Try to get a teenager to change his or her driving behavior only by quoting statistics about traffic fatalities, and you're likely to be met with a blank stare. That's because we humans need more than numbers or the rising curve on a graph to understand the world. We are fundamentally storytellers. Ask the kids in that group of troubled teens about themselves, and they'll respond with a narrative about families and fights, their isolation at school, or the time they ran away or the day a parent skipped out on them. We all use stories to make sense of ourselves in the world. And what's true of individuals is also true of cultures and the sweep of their history.

For most of history, we have used myth to tell our biggest stories. When you hear the word *myth*, you're likely to think of a false story. But, taking a long view of human evolution, myths are often more than just true or false, and they've always played an essential role for us. Every society, in every time and place, has had a system of myths, a constellation of stories that provide a basic sense of meaning and context. Some speak only to our internal life as we make our transitions to adulthood, to parenthood, and to old age. But some tell the big stories. Through these mythic-scale big stories (including their forms in religion), people came to understand how their culture thought the universe was born, how the Earth was formed, and how people were made.

In our age, that role falls to science. Instead of gods and spirits, we now have the Big Bang and Darwin's story of the descent of man. With science, we found a new way to enter into a dialogue with the world, one where experimentation and evidence led the way. That's how the big stories get put together for us moderns. But the power of those stories *as stories* never went away.

When it comes to the fate of our civilization in a climate-changed world, however, we don't have a big story that can convert rising global temperatures and melting Greenland ice sheets into a grand narrative with us in it. The only thing close is a story that goes something along the lines of "we suck." Human beings are greedy and selfish. We are nothing but a plague on the planet.

That story is not only unhelpful and impoverished, it's also entirely wrong from the perspective of the new understanding of life and planets we've recently gained. People often cast the climate crisis in terms of "saving the planet." But as the biologist Lynn Margulis once put it, the Earth "is a tough bitch."[6] It's not the Earth that needs saving. Instead, it's us and our project of civilization that need a new direction. If we fail to make it across the difficult terrain we face, the planet will just move on without us, generating new species in the novel climate states it evolves. The "we suck" narrative makes us villains in a story that, ultimately, has none. What that story does have are *experiments*—the ones that failed and the ones that succeeded.

This larger perspective, gained in light of the stars, does not absolve those who drive climate denial for reasons of greed or political gain. They are fully culpable for their folly. From a planetary perspective and its long view, they will become the reason why Earth's experiment in civilization building fails to reach its higher potential.

So there is an entirely new "big story" we can tell. It's a drama that puts humanity back into the life of the planet. It's a narrative that puts Earth and its life back into the proper context of a universe awash in planets.

In this new story, we aren't collectively villains but we may collectively become losers.

ASTROBIOLOGY AND THE ANTHROPOCENE

Over the last half century, our project of civilization has learned to look out and look back as never before. We have looked back billions of years to uncover the Earth's deep history. We've seen how our species and its civilization comprise just another "expression of the planet," as writer Kim Stanley Robinson calls it.[7]

The evolution of the planet and its life cannot be separated. That is what our science has shown us. Earth and its life must be thought of as a whole that "coevolves" together. Two and a half billion years ago, for example, it was microbes that reworked the world by creating the oxygen-rich atmosphere we now breathe. In the process, those species "polluted" themselves off most of the planet's surface.[8] Using this new oxygen-rich air, the Earth then moved on to create new versions of itself, like the one with the first large sea-creatures, the one with enormous dinosaurs, and the one carpeted by vast grasslands. Those fish, dinosaurs, and grasslands were once new actors on the world's stage. They appeared and took their places in the long drama of Earth's undirected experiments in coevolution. The Earth has run many experiments in life and its possibilities over the last four billion years. We're just the latest version, and in that way we're not so unique.

And just as our science looked back to reveal Earth's history, it also looked outward, traveling across billions of miles to explore the other worlds of our solar system. These audacious journeys showed us that "climate" is not something limited to patterns in your local weather report. On Venus, 220-mile-per-hour winds blow high in the atmosphere.[9] On Mars, icy fog forms each night near its northern pole.[10] There's even rain (made of gasoline) drifting over forty-mile-wide lakes on Titan, the giant moon of Saturn.[11] In terms of having a climate, our planet is also not that unusual.

Finally, we have also looked out to the stars and discovered that the universe is fecund with solar systems like our own. The numbers from these monumental studies tell us that projects of civilization like ours may likely have occurred elsewhere at other points in cosmic history. As long as the universe isn't exceptionally biased against it, we are not the first. Others, on other worlds, will likely have come before us. And with our new knowledge of planets and their laws, we can also see that they too will likely have faced a similar dilemma to the one staring us down today. Even our climate crisis may likely not be so unique and not even that unusual.

So our science has done its job. When it comes to the relationship between life and planets, it has shown us entirely new realities and new possibilities. This science is new, it's revolutionary, and it's called *astrobiology*. Through the truly heroic efforts of scientists across the world, astrobiology has opened up new, universal truths for us about the braided potential for planets and life. It has shown us what *has* happened here, and what *might* happen elsewhere.

That knowledge comes at a strangely auspicious moment, bestowing it with great consequence.

Ten thousand years ago, our project of civilization was born after the beginning of what geologists call the Holocene, a planetary epoch of warm, wet conditions following the end of the ice ages. But in driving climate change, we're now pushing the Earth out of the Holocene into a new era in which human impacts dominate the planet's long-term behavior. The new era is called the Anthropocene.[12]

We all want our project of civilization to continue deep into the Anthropocene. But our efforts so far have mostly failed. We've known about global warming, the most obvious symptom of the emerging Anthropocene, for more than fifty years.[13] Despite having that knowledge, we've done almost nothing to deal with climate change and its consequences. Our politics, our economics, and even our moral philosophy have all failed to drive actions that could ensure the long-term sustainability of our project on a changing planet.

That failure is rooted in the mistaken view that we, and our project, are a one-time story. But we can be forgiven for that failure because, until very recently, we didn't have the tools or the information to rise above such tunnel vision. We did not yet have the astrobiological perspective. But now we do, and it can change the path to our future.

This book explores what might be called the astrobiology of the Anthropocene, and it's built out of two braided questions.

- What can the revolutions of astrobiology tell us about life on other worlds, even other intelligences and their civilizations?
- What can life on other worlds, even other intelligences and their civilizations, tell us about our own fate?

These interwoven questions will lead us to a fundamentally new story about what we are and what's happening to us at this crucial moment in our civilization building. It's a narrative built from space telescopes, deep-sea submersibles, robots diving into comets and geologists scrambling over deadly glacial chasms. In telling that story we'll encounter science that is nothing less than thrilling.

The Astrobiology of the Anthropocene will show us images of steep cliffs under coral skies on Mars that help to vastly enlarge our understanding of climate and climate change. It will take us to dark crazy-quilt ecosystems deep in the ocean that give us a time-machine view of Earth billions of years ago when life was still new.

And then there are all the new planets.

The astrobiology of the Anthropocene will also take us across the galaxy to see entirely unexpected classes of planets have just been added to the textbooks: "hot" worlds snuggled tightly against their parent stars and "super-Earths" many times the size of our own.[14]

In telling this new story, we will also encounter the most thrilling of all possibilities: *aliens*.

Our exploration of the astrobiology of the Anthropocene will lead us to a radical claim. It's time to take the existence of aliens—by which we really mean *exo-civilizations*—seriously. Everything that has been learned in the astrobiological revolutions of the last few decades now allows us to see just how improbable it is for us to be the only project of civilization in cosmic history. That realization tells us that if we ask the right kinds of questions, the ones backed by the hard numbers of the new exoplanet discoveries, we can begin making out the contours of a science of exo-civilizations that's relevant to our own crisis on Earth.

The new science this book explores won't tell us if the galaxy is teeming with other civilizations now. It won't tell us if we're going to catch evidence of their existence soon. It won't tell us if they have pointy ears or seven-fingered hands or are shaped like lizards. What it can do, however, is show us how *this*—meaning everything you see around you in our project of civilization—has quite likely happened thousands, millions, or even trillions of times before.

From the vantage point of the astrobiological data, we can take those exo-civilizations seriously as a subject of scientific inquiry. It's hard to avoid the giggle factor when we talk about aliens. Years of bad TV science fiction (as well as the crazy world of UFO conspiracy theorists), have left a bad taste for most scientists wanting to think about intelligent life on other worlds. Worse still, for years there wasn't much in the way of scientific constraints on the question. Without such constraints, the discussion falls dangerously close to pure fiction. But if we ask the right kind of questions, the laws of planetary behavior we have now grasped can act as guardrails for how we think about exo-civilizations. That means if we ask the right questions, we can get answers. At the very least, with the right questions, those laws of planets we've uncovered will help us put limits on what answers can look like.

Remarkably, the one domain where such questions exist lies exactly at the intersection of astrobiology and the Anthropocene. Our new understanding of planetary laws is well tuned to address the question we most care about: How will a civilization (meaning *any* civilization) coevolve with its planet (meaning *any* planet)? If other civilizations have likely existed across cosmic time and space, we can take them seriously by making them the subject of our science. We can bring to bear all we've learned from Earth, Venus, Mars, and the thousands of planets discovered outside our solar system. We can deploy the laws of physics and chemistry inherent in that knowledge to begin doing science in the form of detailed models and simulations.

From this perspective, civilizations become just another thing the universe does, like solar flares or comets or black holes. We can use what the stars have laid out before us in our astrobiological studies to explore how any civilization on any planet can—or, in the worst case, cannot—evolve together. We can treat those possible exo-civilizations as other histories that can tell us about our own future.

The advantages of this astrobiological perspective can be gained even if no other civilization ever existed. Thinking about hypothetical exo-civilizations is valuable in dealing with the challenge of the Anthropocene because it teaches us to "think like a planet." It teaches us to frame our pathways to a long-term project of civilization in terms of the coevolution between life (including us) and the Earth. Through the astrobiological perspective, we can map out the contours of our own fate and our own future.

A NEW STORY

If, however, we take the possibility of other civilizations seriously, then we find a new door open to us in facing the Anthropocene. Across a cosmos with so many planets experimenting with life, we can see that some technological species may learn how to make it. They will learn how to navigate the difficult bottleneck of the climate feedback they generate. Others, however, will fail. That's how this new big story becomes meaningful. It begins with the science, but ends with showing us how to face the hard choices our version of the Anthropocene forces on us. A long-term sustainable version of our project of civilization will mean we must become partners, in some as-yet-unknown way, with the planet.

We must therefore become a power—in our own right—alongside the Earth. But as Spider-Man so famously discovered, with great power comes great responsibility. Does becoming a winner in the game of cosmic evolution mean we hold the Earth in a perpetual version of the Holocene? Will we never allow another ice age to form? If that's true, then what about the species that might have emerged in the ice ages we block? Do we have the right to keep them from ever entering the Earth's drama?

And which species from the current version of the Holocene do we carry with us into the Anthropocene? Images of polar bears adrift on lonely ice floes tug at our hearts. But entering into a true, long-term partnership with the planet will demand hard choices. Those decisions will not just be the domains of science. They will also depend on what we value, what we hold dear, and what we believe to be sacred. These are all the domains of meaning. That is why getting the story right—the story with us in it—is just as important now as getting the science right.

The astrobiological perspective on the Anthropocene is science at the grandest of scales. It's a narrative of our own collective life and fate set against the stars, whose own stories suddenly matter as both our guides and our teachers. It is more than data, more than information, more than knowledge. We, and our cherished project of civilization, are crossing over a frontier as our planet enters the Anthropocene. The new science we'll explore in what follows can help us map this new territory. It can also help us to navigate its burning edges and make it through to the other side.

CHAPTER 1

THE ALIEN EQUATION

THE FERMI PARADOX

One warm and bright summer day in 1950, four colleagues walked through the atomic weapons complex at the Los Alamos National Laboratory in the high desert of northern New Mexico. The Cold War with the Russians was in full swing, and there were new faces everywhere. Each of the men, however, was an old hand at the lab, and each played a key role in developing the bombs that helped win World War II.

First among them was Enrico Fermi, the Italian-born Nobel laureate whose brilliance had pierced the mystery of the atomic nucleus. Fermi was famous for his almost superhuman scientific abilities. C.P. Snow once wrote that if Fermi had been born just a bit earlier, he might have invented all of atomic science by himself. "If [that] sounds like hyperbole," Snow wrote, "anything about Fermi is likely to sound like hyperbole."[1]

Walking alongside Fermi was Edward Teller, the brooding Hungarian physicist whose work would become synonymous with the terrifying hydrogen bomb. While Fermi was not in favor of Teller's push for the "super" bomb, the two men remained friends throughout their lives.[2] Rounding out the group that day were American nuclear scientists Emil Jan Konopinski and Herbert York, both highly regarded researchers in their own right.

The four scientists made their way from the lab buildings to the Fuller Lodge where lunch was served (it was one of the few structures left over from the site's earlier incarnation as a boys' camp). As they walked, the conversation turned to unidentified flying objects.[3] Since the end of the war, sightings of mysterious lights in the sky had been increasing. A recent incident had just made the local papers, reminding York of a whimsical *New Yorker* cartoon in which flying saucers were blamed for a rash of Manhattan garbage can disappearances. Given the physicists' inclination for analysis, the UFO story led to a tussle of questions about faster-than-light travel and its limitations. Soon, however, the conversation wandered off to

Alan Dunn's cartoon of UFOs abducting New York City garbage cans, which appeared in *The New Yorker* in 1950.

other topics as the four scientists continued along the path lined by pine trees and juniper. It was only later, in the middle of lunch, that Enrico Fermi blurted out, "But where are they?"[4]

Teller, York, and Konopinski all broke into laughter at Fermi's outburst. They recognized their colleague's sharp insight. Fermi had a habit of reducing complex problems to the barest essentials. Present at Trinity, the desert test of the first atomic bomb, Fermi had famously calculated the explosion's power by simply dropping scraps of paper and noting how far sidewise they were swept by winds from the blast.[5]

But on that summer day over lunch, Fermi had identified a core question destined to haunt all subsequent discussions of intelligent life in the cosmos. Fermi's observation was as straightforward as it was penetrating. If the evolution of extraterrestrial intelligent species was common, why didn't we see them? Why hadn't our telescopes found indications of their existence? Why hadn't aliens already landed on the White House lawn?

Fermi's question was not aimed at UFOs. That topic was, and continues to be, a morass of weak reasoning, poor observations, fakery, and conspiracy theories. Instead, his question would come to represent one of the first distinctly modern and scientifically manageable questions about extraterrestrial technological civilizations (we'll use the term *exo-civilizations*).[6]

Over time, Fermi's question would come to be known as Fermi's Paradox. Its formal statement might go as follows: If technologically advanced exo-civilizations are common, then we should already have evidence of their existence either through direct or indirect means.

In the decades to come, other scientists would give Fermi's question the precision it needed to have a scientific bite. In 1975, astro-physicist Michael Hart's paper "An Explanation for the Absence of Extraterrestrials on Earth" addressed a number of objections to the reasoning behind Fermi's paradox, including ones associated with physics, biology, and sociology.[7] His conclusion was that none of the objections was strong enough to put off the paradox's logic. Hart laid bare the essence of Fermi's insight by demonstrating that just one species could "quickly" colonize the galaxy. Assuming an exo-civilization appeared that built ships capable of traveling at 10 percent of the speed of light, Hart showed that within just 650,000 years these creatures would cross the width of the galaxy. In this way, a single species could send ships in all directions, radiating out-ward from their home world, and quickly colonize every star system.

Of course, a few million years seems like a long time to most of us. Our species, *Homo sapiens*, has been around on the Earth for less than a million years. But what's long for us is short for the life of the galaxy. The Milky Way, our home galaxy, is a vast and ancient metropolis of stars. It was born some ten billion years ago. So it would take about one ten-thousandth of the Milky Way's age for Hart's spacefaring civilization to cross the galaxy. Hart had demon-strated that, in a time scale that is small compared to the galaxy's existence, just one randomly appearing interstellar civilization could reach all the planets orbiting all the stars in the sky—*including our own.*

For some researchers, Hart's work filled the night with a disqui-eting emptiness. In their eyes, there was a straightforward logic to the Fermi Paradox that said we must be alone. The obvious absence of an alien civilization in our solar system, along with the lack of evi-dence for those civilizations' existence among the stars, must mean no other form of life anywhere had reached our level of intelligence and technology. We were the sole species in the Milky Way that had made it up the ladder of evolution to build an advanced civilization. In response to the Fermi Paradox, physicist and science fiction writer David Brin spoke of the stars' "Great Silence." It was an apt term, capturing the cosmic loneliness that Fermi's Paradox seemed to imply.[8]

Along with the Great Silence, an increasing interest in Fermi's Paradox led to the idea of a "Great Filter."[9] The absence of evidence for advanced civilizations in the galaxy does not imply that Earth is the only life-bearing planet. The Fermi Paradox only speaks to the existence of technological civilizations like ours, or ones even more advanced. Microbes or shellfish or even dinosaurs might exist on every world in the cosmos. So if we don't see exo-civilizations, some scientists argued, there must be a filter keeping evolution from spawning them. In other words, if we are alone in the cosmos, then some kind of evolutionary wall blocks other planets from reaching our level.

But a Great Filter might lie anywhere along that evolutionary path. Perhaps simple life is so difficult to form that *it* constitutes the Great Filter. In that case, Earth would be one of the few worlds with life. On the other hand, the emergence of even simple forms of intelligence might be the Great Filter. So, while lizards might appear on many worlds, dolphins and apes would not. If that were true, the difficulty in evolving intelligence filters out even those worlds where life *has* formed from moving further toward a technological civilization.

Ironically, at the exact historical moment that Fermi and his colleagues were sitting down to lunch, a new kind of evolutionary dead end for the Great Filter made its appearance. Fermi posed his question at a laboratory dedicated to developing weapons of unprecedented destructive energies. It was in the 1950s that humanity first gained the power to bring civilization to an abrupt and decisive end through full-scale nuclear war.

Atomic Armageddon made it possible to imagine that the Great Filter lay not in the distant evolutionary past (in which case we had been lucky to avoid it); instead, it might wait like a viper, hiding in the tall weeds of our future. Maybe the night sky was silent—and our planet unvisited—because no advanced civilization was smart enough to handle the pressure of its own existence.

If someone could have asked Fermi for his top choice for a Great Filter, he would likely have answered nuclear war. These days, however, we have a broader understanding of civilizations and their existential challenges. In the 1950s, when Fermi posed his question, there was only a small community of Earth scientists awakening to the possibility of human-driven climate change. The idea that humans could unintentionally change the behavior of the entire planet through nothing more than our collective daily activity was an idea so radical, it had barely been formulated in a scientific way. Now, however, we know better.

Earth's passage into its human-dominated era, increasingly known as the Anthropocene, shows us a potentially more potent candidate for the Great Filter. Civilizations like our own are a complex web of interdependent systems. Where would you get your food if the electricity went out for a year? How would you heat your home if the pipelines delivering petrochemicals shut down? There are a million ways we all rely on the smooth operation of these systems.

But a significant shift in Earth's climate state would upend those systems in ways that would deeply challenge their operation.

Think about the Gulf Stream for a moment. It cycles warm water (and warm weather) up from Florida to Boston, and then out across the Atlantic. Hundreds of millions of people in some of Earth's most technologically advanced cities rely on the mild climate delivered by the Gulf Stream. But the Gulf Stream is nothing more than a particular circulation pattern formed during a particular climate state the Earth settled into after the last ice age ended. It is not a permanent fixture of the planet. If the climate changes enough, the Gulf Stream, and the mild weather it delivers, could become a thing of the past.[10]

So what we call the Anthropocene may be a far more potent candidate for the Great Filter than nuclear war. An all-out nuclear exchange would, after all, be intentional. It would be someone's decision. But it's easy to imagine other civilizations less aggressive and warlike than ours. They might not even think to build nuclear weapons. Climate change, however, is likely to be universal. As we will see, it is likely to be a consequence of any project of advanced civilization building on any planet. Long-term dramatic climate change need not lead to a civilization-building species' extinction. It only needs to make conditions difficult enough that their project of technological civilization is disrupted and unable to recover on the now-climate-changed planet.[11]

All these issues surrounding the Great Filter really illustrate the power of Fermi's insight. Making progress in science often hinges on asking the right kind of question. Without a well-posed question, discussions become little more than people talking (or yelling) past each other. And without a well-posed question, there's no clear path toward gathering data that will yield answers.

Finding a good question is like throwing open the shades in a dark room. It's the first step in finding a new way to tell a story about the world because it lets us see the world in a new way. A good question reframes what we think is important. It tells us where we should be looking, where we should be going, and how to begin organizing our efforts to get there.

Fermi's 1950 question helped play that role for the issue of exo-civilizations. As developed by Hart and others, Fermi's Paradox asks us to consider if and why humanity might be alone in the universe.

But to truly understand the importance of Fermi's question for our future, we need to travel back a few thousand years into our past.

THE PLURALITY OF WORLDS

The Greek philosopher Epicurus made the first expression of what we might call "exo-civilization optimism" almost 2,200 years ago: "There are infinite worlds both like and unlike our own. . . . Furthermore we must believe that in all worlds there are living creatures and plants and other things we see in this world."[12]

Epicurus's interests ranged from ethics to the nature of suffering, but first and foremost he was an atomist. The world for him was composed of an infinity of tiny components, arrayed in infinite combinations. That belief served as a foundation for the atomists' belief that the universe must also be infinite, and thus must contain an infinite number of other inhabited planets.

Not all Greek philosophers, however, shared the atomists' faith in a fecund cosmos. "There can not be several worlds," wrote Aristotle in nearly the same era.[13] Aristotle was an exo-civilization pessimist. For him, the Earth was the center of the entire universe. Since there can only be one center, the Earth must be unique. Aristotle was certain that no other worlds, and certainly no other worlds like Earth, existed.

The conflict between these convictions—of fecundity of the universe on one hand and the uniqueness of the Earth on the other— would echo down the next twenty centuries. From the Greeks, through the Middle Ages, to the Renaissance, and on into the early twentieth century, optimism concerning other inhabited planets waxed and waned.

From one century to the next, philosophers, physicists, theologians, and astronomers asked the same questions: Are we alone? Are we the first? Each generation posed the question using the prejudices, ideas, and tools of their time. The arguments were always fierce; sometimes they even turned deadly. In the medieval period, the Catholic Church considered discussion of other worlds to be heresy. That did not stop philosophers and theologians from struggling to understand why an infinitely powerful God would create only a single inhabited world. In the thirteenth century, Thomas Aquinas answered this dilemma by claiming God could have created other inhabited planets, but had chosen not to (a distinctly unsatisfying solution).[14]

By the sixteenth century, a new generation of thinkers was pushing back on the question of other worlds. Copernicus famously dethroned the Earth from the center of the universe in *On the Revolution of Heavenly Spheres*, first published in 1543. In his version of astronomy, radical for its time, our world was just one more planet orbiting the Sun.[15] Copernicus never expressed opinions about other planets orbiting other stars. But his work removed Earth from its privileged cosmic position and opened the door for others to publicly explore what became known as "the plurality of worlds" question.

For a time, the Church tolerated some discussion of Copernican astronomy. But in the late 1500s the radical Dominican monk Giordano Bruno pushed the limits of that tolerance until it broke. Bruno not only publicly advocated for Copernican astronomy, he went further, arguing that the universe must contain infinite worlds with infinite varieties of inhabitants. These views helped earn him the attention of the Inquisition, and in 1600 the Church burnt Bruno at the stake for heresy.[16]

By the time the scientific revolution was in full swing, Isaac Newton had revealed powerful, unifying laws governing the motion of celestial *and* terrestrial objects. Astronomy was making swift progress, as new planets such as Uranus and Neptune were discovered and the orbits of comets were understood. The intellectual tumult shifted debate about life on other worlds for both scientists and an increasingly literate public. The influential French writer Bernard de Fontenelle, for example, scored the equivalent of an Enlightenment-era best seller with his 1686 book *Conversations on the Plurality of Worlds.*

The book was framed as a series of late-night discussions between a philosopher and a quick-minded young baroness. Expressing the optimism of his age, de Fontenelle imagined that many of the planets orbiting the Sun hosted peoples. He even thought the Moon had intelligent inhabitants. Turning his sights beyond our solar system, de Fontenelle wrote, "The fixed stars are so many Suns, every one of which gives Light to a World." And on many of these worlds, de Fontenelle was certain that life thrived.[17] One influential image from the book gives a graphic representation of de Fontenelle's optimism. The frontispiece of an early edition shows our solar system nestled snuggly amidst a cosmos dense with other stars and other worlds.

It was an optimism that prevailed well into the nineteenth century. Darwin's theory of evolution brought a new twist to the dis-

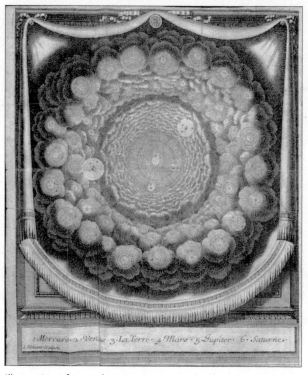

Illustration of our solar system surrounded by other stars and their planets from Bernard de Fontenelle's 1686 book, *Conversations on the Plurality of Worlds*.

cussion of life and planets. Writers like Camille Flammarion, the French Carl Sagan of his day, thrilled audiences with visions of life evolving in entirely novel forms on a fertile Mars and Venus.[18] Adding evolution theory to debates about the plurality of worlds gave writers like Flammarion the chance to imagine how nature shaped the inhabitants of other planets. Since evolution responded to the specific conditions on a given planet, the transformations a species undergoes must fit those conditions. In this way, Flammarion argued that life on Mars must be very similar to life on Earth, since both planets (he thought) presented similar environments.[19]

Mars would later become the focus of a very public version of optimism. At the turn of the twentieth century, American millionaire Percival Lowell founded an observatory in Flagstaff, Arizona (which had yet to achieve statehood and was still a territory), to study so-called "canals" on the Red Planet.[20] Lowell was convinced that Mars was inhabited. Through books and lectures, he dedicated his final years to convincing others. His efforts were successful enough that many in the general public took it as a given that Mars was a living world.

During the latter half of the nineteenth century, however, a pessimistic pushback on exo-civilizations emerged, both from outside and within science. In 1853, William Whewell, an English scientist, philosopher, and Anglican priest, wrote a scathing critique of the optimists' position in his book *Of the Plurality of Worlds.* Turning from mere hopes expressed by other writers to the astronomical facts of his day, Whewell wrote, "No planet, nor anything which can fairly be regarded as indicating the existence of a planet revolving about a star, has anywhere been discovered."[21] Whewell also argued strongly against using Earth's history as a guide for life's progress on other worlds. "The assumption that there is anything of the nature of a regular law or order of progress from [interstellar] material to conscious life . . . is in the highest degree precarious and unsupported."[22]

Another dissenting voice came from Alfred Russel Wallace, who, along with Darwin, is considered one of the founders of evolution theory. In his 1904 book *Man's Place in the Universe*, Wallace applied his own detailed understanding of biology to the question of life on other worlds. Using the availability of liquid water as a guide, Wallace concluded Earth was the only habitable solar system world. Going further, he claimed few planets in the galaxy would be earthlike enough to allow for intelligence.[23]

By the early twentieth century, a more determined pessimism about the existence of planets around other stars (now called "exoplanets") took hold. It was a view that proved damning for scientific views of exo-civilizations as well. This new pessimism focused on the prevalence of planets and rested on a model for planet formation called *collision theory*. Theoretical studies by astronomers in the early 1900s argued that planets could only form when two stars passed in a close encounter. As the suns shot past each other in a near collision, gravity would pull some of their gas into space, leaving it to fall into orbit around one of the stars. Eventually, the extruded gas would cool and coalesce into a planet. James Jeans, the leading astronomer of his day, soon demonstrated that these kinds of stellar near misses would be exceedingly rare. Because of Jeans's work, by the middle of the twentieth century, many astronomers believed planets were few and far between in the universe.[24] That meant life would also be rare.

So, by the time Fermi and his companions sat down to lunch

that day in 1950, the buoyant optimism of de Fontenelle and Flammarion had been stalled. Many scientists thought planets were rare. Even if they weren't, biological arguments like those of Alfred Wallace could be marshaled to make life seem like an improbable event. Even worse for those who wanted to take life on other worlds seriously, Lowell's observations of Martian canals had become a joke in the scientific community.[25] In the early 1950s, the possibility of life and intelligence in the universe remained a question that few scientists were seriously considering.

But science does not exist in a vacuum. It is a human endeavor, and its story evolves with the stories laid out by the rest of a culture, even as it shapes that culture. The narratives we could tell ourselves about life in space were set to shift for the worst of reasons.

ROCKETS, BOMBS, AND SATELLITES

When Fermi blurted out his famous question in 1950, the US was still reeling from news that Russia had detonated its own atomic weapon. At that time, the total US inventory of atomic bombs numbered in the hundreds. By 1960, however, the global weapons stockpile had grown to more than twenty-two thousand.[26] More importantly, the early bombs had been based on nuclear fission—the splitting of the nucleus of a heavy atom like uranium. The carnage at Hiroshima and Nagasaki had demonstrated that these "atomic" weapons could wipe out a large portion of a city in an instant. By 1960, both the US and Russia had deployed weapons based on thermonuclear *fusion*. These bombs were powered by slamming atoms of hydrogen, the simplest element, together to create something heavier, following the same basic process that powers stars like the Sun. The new hydrogen weapons were terrifyingly powerful.[27] A medium-sized H-bomb could destroy an entire metropolitan area. The largest H-bomb could blow a small portion of the Earth's atmosphere into space.

The race toward ever more powerful nuclear weapons defined much of the 1950s. But the bombs triggered another race during that decade, and this second technological sprint would have an even greater impact on reimagining the fate of far-flung exo-civilizations.

Building more powerful bombs meant little to nuclear weaponeers if they couldn't be delivered to their targets more quickly than those of the enemy. In this way, the logic of the Cold War moved inexorably from the technologies of jet bombers to those of rocket-powered missiles.

In the final years of World War II, Nazi V-2 guided missiles had terrorized Britain and proven the power of long-range rockets. After the war, both the Russians and the US snapped up captured German V-2 scientists, and each nation vigorously pursued the development of continent-crossing rockets called intercontinental ballistic missiles (ICBMs). The Russians proved faster and more nimble in their development. On August 21, 1957, a Soviet R-7 missile blasted across 3,700 miles, reaching an altitude of ten miles.[28]

The true power of these rockets became apparent two months later, when the world woke up to find we'd acquired a second moon. On October 4, 1957, another Russian R-7 rocket punched the 184-pound *Sputnik* above the Earth's atmosphere and into orbit, where it became the Earth's first artificial satellite. Wheeling a few hundred miles overhead, *Sputnik* broadcast perfectly timed radio "beeps" for anyone with the right equipment to hear.[29] And the world *was* listening. While Russian politicians gloated and Americans panicked, it was clear that an ancient threshold had been breached. Humanity's space age had begun.

There was, however, only one way to talk to a hypersonic rocket in the atmosphere or a satellite orbiting high above the planet. Communications at these ranges required the use of sophisticated radio technology. And it was exactly in those technologies that the political and military urgencies of the 1950s dovetailed with the first scientific effort to detect alien intelligence.

Until the 1950s, astronomy was carried out with telescopes fashioned with glass lenses and mirrors. That meant astronomy was done only with visible light—the kind our eyes had evolved to perceive. But visible light is nothing more than waves of electromagnetic energy with wavelengths that fall within a certain range. (Wavelength is the distance between the peaks in a wave.)

In the mid-1800s, physicists discovered there was an entire spectrum of electromagnetic waves. These waves stretch from very short, atomic-scale X-rays and gamma rays all the way to radio waves the size of buildings. Astronomical objects tend to emit energy across a large fraction of this electromagnetic spectrum.

Evolution tuned our eyes to see electromagnetic waves only in the visible "band" of the spectrum. It's no coincidence that this visible band happens to be where the atmosphere is most transparent to sunlight. But the Sun also produces X-ray "light," ultraviolet "light," and radio "light."

Buoyed by the advances in radio engineering during World War II, astronomers in the 1950s began opening their first new "window" on the night sky by using light outside the spectrum's visible band. With radio waves, researchers found they could map out the entire galaxy or capture the echo of long-dead stars in ways that were impossible using visible light.

Radio astronomy, as it was called, constituted one of the most exciting frontiers in science as the 1950s progressed. If you were young, gifted, and scientifically ambitious, radio astronomy was the place to be. That was how, at the end of the decade, a newly minted astronomer named Frank Drake found himself in the wilds of West Virginia, searching for signals of alien civilizations.

LISTENING TO THE SKY

Frank Drake had always been a gearhead with vision. The man who would help define much of the modern science around exo-civilizations was born in 1930 on the south side of Chicago, just as the Great Depression began. His father, a chemical engineer for the city, often brought home gadgets for his son that ended up in the boy's basement "lab." The young Drake spent hours in that basement, playing with motors, radios, and chemistry sets. But it was the frequent bike trips to the city's Museum of Science and Industry that took Drake's imagination beyond the details of his radios. There, he and a friend found full-scale models of atoms that made the invisible real. "Some of the exhibits were so dramatic, it would almost knock you to the floor," Drake later wrote.[30]

When Drake was just eight years old, his father told him there were other worlds "just like Earth." The idea gave him a vision of other life and other planets that never faded. The Oz stories were also a favorite of young Drake. As a child, he owned many of these books about another world. Author L. Frank Baum had written thirteen volumes beyond the first one, *The Wonderful Wizard of Oz*, many of them featuring Princess Ozma, the ruler of Oz.[31]

The boy grew into a tall, handsome young man with an affinity for science that landed him at Cornell University with a Reserve Officers' Training Corps scholarship. While Drake didn't begin his undergraduate work with a specific interest in astronomy, he soon found himself drawn to the subject. And throughout his introductory astrophysics courses, he never lost his fascination with the question his father introduced to him as a boy: Are there other inhabited worlds in the universe? But it was not a question he was willing to pose to his professors, for fear of sounding like a fool. That reticence would fade through a chance encounter with Otto Struve, one of the world's most famous astrophysicists.

Struve was a large, intimidating man who was a leader in the study of stars. In 1951, he was invited to present a lecture to the Cornell community, and Drake was in attendance. The lecture focused on what was known about how stars formed from clouds of interstellar gas. As he neared the end of his talk, the imposing Russian-American pivoted to the topic of life in its cosmic context. He claimed there was mounting evidence that at least half of the stars in the galaxy had their own planetary systems. The old collision theory of planet formation was falling from favor, and Struve said there was no reason why life couldn't exist on some of those planets.[32] A light went on in Frank Drake's head. Here was someone older and established, asking the same question he'd been fascinated by since he was a boy.

Struve's inspiration was still alive in Drake in the spring of

1958 as he piloted an old white Ford, stuffed with all his belongings, through the backwoods of rural West Virginia. He was on his way to the newly minted National Radio Astronomical Observatory's Green Bank facility. There, he was to become a member of the observatory's fledgling scientific staff.

The research engines of the Cold War were churning, and funding had been opened for any project that could push American capabilities forward. In Drake's words, Green Bank "had been given what amounted to unlimited funds to build the best radio observatory in the world."[33] Nestled in a remote, verdant valley valued for its radio (and actual) isolation, Green Band was the new home of American radio astronomy.

Soon after Drake's arrival, the towering metalwork of an eighty-five-foot radio dish was completed. The astronomers at Green Bank planned to use the newly commissioned telescope to study everything from the structure of our galaxy's pinwheel shape to its hidden center.[34] Drake would be part of many of these efforts. But the inhabited worlds in Drake's imagination wouldn't leave him alone. It wasn't long before he was thinking of ways to use this giant radio ear to find them.

"I calculated just how far our new 85-foot telescope could detect radio signals from another world if they were equal to the strongest signals [on earth]," Drake later wrote.[35] The answer turned out to be about ten light-years, or sixty trillion miles. Since he believed stars like the Sun had the best chance of hosting a world like Earth, his next step was to check the star charts. Luckily, there were at least few sunlike stars within ten light-years.[36] Drake saw he had the beginnings of a real research project.

After his initial calculation, Drake needed to get his colleagues

Frank Drake and the early telescope at the
National Radio Astronomy Observatory in
Green Bank, West Virginia, in 1964.

at the observatory to buy into something as seemingly crazy as a search for alien civilizations. The scientists who lived at Green Bank often ate together at a roadside diner a few miles away. Over lunch there one winter day, Drake made his pitch to use the telescope to search for signs of intelligent life on other worlds.

"At the time, the director of the National Radio Astronomy Observatory was Lloyd Berkner, [who was] something of a scientific gambler, and he was all for it. So as the last greasy french fry was washed down by the last drop of Coke, Project Ozma was born."

"Project Ozma." True to the exuberance of his childhood dreams, Drake named his search after the princess of the Emerald City. With the blessings of the observatory administration, the team began building the equipment needed to carry out Project Ozma. By the spring of 1960, the amplifiers, filters, and other radio engineering gear were ready.[37]

For six hours each day that year, from April to July, Drake aimed the telescope at one of two target stars. The first was Tau Ceti in the constellation Cetus (the Whale). The second was Epsilon Eridani in the constellation Eridanus (the River).[38]

He later wrote of remembering "the battle against the cold each morning as I would climb to the focus of the dish. . . . And then of that moment on the first day of the search when a strong, pulsed signal came booming into the telescope just as soon as we had turned it towards Epsilon Eridani."[39]

The heart-pounding excitement of the "booming signal" turned out to be a false alarm. That source turned out to be man-made. It was, in fact, just about the only time Drake thought they'd detected another civilization. Project Ozma never captured any alien signals, but it did capture something else of great importance: the world's imagination.[40] Just ten years after Fermi had asked his question among a small group of friends, at least some in the scientific community were ready to take the question of exo-civilizations seriously.

As Drake was working out the details of his search at Green Bank, two physicists named Giuseppe Cocconi and Philip Morrison published a groundbreaking study titled "Searching for Interstellar Signals." The paper appeared in a 1959 issue of *Nature*, one of the most prestigious journals in science. The two physicists argued that the best way to look for signals from advanced exo-civilizations was by using radio astronomy. Dust blocks visible light, making the Milky Way seem blotchy to our eyes. But radio "light" has long wavelengths that pass unobstructed through dusty regions of the galaxy. So, with radio waves, the galaxy becomes transparent, allowing astronomers to "see" from one end to the other. This meant a civilization emitting radio waves could be seen at far greater distances than one emitting visible light signals.[41]

Drake had already reached the same conclusion. But the publication of Cocconi and Morrison's paper meant others were thinking exactly along his lines. It was a development that worried the new director of Green Bank, who was none other than Drake's inspiration, Otto Struve. Until then, Drake had been keeping a tight lid on his search. Struve, however, feared getting scooped. Within a few weeks, Struve used an invited lecture at MIT as an opportunity to reveal Project Ozma's existence to the world.[42]

Soon, Drake was hosting a steady stream of visitors. Award-winning journalists, theologians, and leading businessmen made the trek to Green Bank. Project Ozma, along with the publication of Cocconi and Morrison's paper, marked a turning point in the way science engaged with the issue of alien civilizations. By 1960, humanity was dogged by questions of its own imminent destruction on one hand, while it watched the space age dawn, offering fresh possibility, on the other. These two technological developments were reshaping politics and culture, and they served as a kind of imaginative ether, launching the first true scientific search for other civilizations.

With Project Ozma, a specific scientific question about exo-civilizations had finally been posed in a way that could be explored using a specific set of appropriate scientific tools. As this crucial threshold was crossed, exo-civilizations rose for the first time from the purely speculative realm of science fiction. One year later, the young Frank Drake would see the consequences of this work become manifest in a fateful call from Washington, D.C.

THE GREEN BANK CONFERENCE

J. Peter Pearman was a staff officer of British origin at the National Academy of Sciences. In the summer of 1961, he called Drake with a remarkable request. Pearman was part of the Academy's Space Science Board, and he wanted Drake to host a meeting exploring the research possibilities for "extraterrestrial communications." Drake had spent the year after Project Ozma nervously wondering which of his colleagues might be snickering behind his back. He agreed immediately to run the meeting.[43]

The discussion then turned to invitations. Drake was happy to discover from Pearman that not only were other scientists taking up the question of extraterrestrial life, but there were two government-sponsored committees already exploring the problem. Together, they drew up a list of ten scientists for the meeting.

First, there would be Cocconi and Morrison, the authors of the *Nature* paper. Drake then suggested Dana Atchley, a radio engineer who'd donated a key piece of equipment for Project Ozma. Barney Oliver, a Hewlett-Packard "research magnate" who'd visited Drake during Ozma, was also included. As a leading astronomer and head of Green Bank, Otto Struve was asked to serve as the meeting's chairperson. Struve then asked that his former student Su-Shu Huang join the group. For expertise in the chemistry of life, the pair chose Melvin Calvin, a Berkeley scientist who discovered the chemical pathways of photosynthesis that allow plants to turn sunlight into food. Rumors were flying that the next Nobel Prize in chemistry would have Calvin's name on it.

Running over their list, Drake joked, "We've got astrophysicists, astronomers, electronics inventors, and exobiology experts. All we need now is someone who's actually spoken to an extraterrestrial."[44] Without missing a beat, Pearman, in his perfect Oxford accent, told Drake he had exactly that. John C. Lilly was a biologist who had become famous for his work with dolphins. Lilly claimed his research demonstrated that dolphins were as intelligent as people. Lilly also believed they possessed a sophisticated form of language that he could decipher. Drake agreed that Lilly should be on the list.

There was one more scientist Pearlman and Drake wanted to invite. He was younger than all the other invitees, but his name, like Drake's, would shape the future of astrobiology. In the summer of 1961, Carl Sagan was newly minted PhD with a fellowship at Berkeley. There, he'd been working with Calvin, developing laboratory experiments on the formation of life. Though only twenty-seven, Sagan had already made a name for himself as both brilliant and brash.[45]

The meeting was scheduled for October 31, 1961. Invitations were sent out, and Drake and Pearman were soon delighted to find that almost all were accepted. Only Cocconi declined (he would never engage in astrobiological research again). But as the meeting approached, a conflict appeared. The group had gotten word that Calvin was going to get his Nobel Prize in chemistry, and the announcement would come during the three days of the Green Bank meeting. Calvin was more than willing to take the call from Sweden at Green Bank, but Pearman and Drake knew some champagne would be needed for a celebration. Procuring bubbly, however, posed its own kind of challenge.

"[Getting champagne was] no mean feat in the semidry state of West Virginia," Drake later recalled. "West Virginia apportioned one state-operated liquor store to each county. The one closest to the observatory stood in a little lumber town called Cass, about ten miles away. The observatory's staff now included a driver—a West Virginian with the fairly common (for those parts) first name of French, and the improbable surname of Beverage. For a moment I considered sending him to buy the champagne, but it would have been too silly. Instead, I drove over to Cass myself that weekend."[46]

Drake purchased a case of champagne and made his way back to Green Bank.

With the invitations complete and the champagne hidden away, the only thing left for Frank Drake to do was to set an agenda. "I sat down and thought, 'What do we need to know about to discover life in space?'"[47]

Drake simply wanted a way to organize the discussion, but the path he chose had consequences far beyond the Green Bank conference. Though Drake could not have known it at the time, his idea would establish an organizing principle for the entire future of astrobiological science.

Since the purpose of the meeting was to explore possibilities for communication with exo-civilizations, Drake understood that the first and most important question would be how many exo-civilizations there were to communicate with. That translated into a single, specific question the meeting needed to answer: What is the number of technologically advanced civilizations in the galaxy that can emit radio signals detectable on Earth?

The galaxy contains about four hundred billion stars.[48] If the number of technological civilizations (call the number N) turned out to be small, then the search for exo-civilization would be unlikely to succeed. There would be just too many stars to search and too few inhabited systems to find. But if N were large (in the billions, perhaps), then astronomers wouldn't have to search many stars before an exo-civilization popped up.

So, what Drake needed was a way to estimate the value of N. To accomplish this, he broke the problem up into seven pieces. Each piece represented a distinct subproblem the scientists at the meeting could discuss in detail. Most importantly, each piece could be expressed as a factor in an equation for the number of galactic exo-civilizations—the all-important quantity N.

Let's run down the seven pieces of Drake's equation and his exo-civilization question.

1. The Birth Rate of Stars

Based on our own experience here on Earth, life will form on planets. Of course, it is perfectly reasonable to ask whether life can bypass planets by forming in something like an interstellar cloud (astronomer Fred Hoyle assumed this in his famous sci-fi story *The Black Cloud*).[49] Given what we *do* understand about the mechanisms of life, however, it's far more likely that a solid planetary surface with lots of liquid water and other chemicals is a requirement to get biology going. Assuming a focus on planets brings us straight to a focus on stars. If we want to know how many planets host exo-civilizations in the galaxy, we first have to know how many planets exist, and that means we first have to know how many stars exist.[50] So Drake's equation begins with the number of stars created in the galaxy each year. Astronomers represent this by the symbol N_* (read as "N sub star").

2. The Fraction of Stars with Planets

Once we know the number of stars forming per year, we can then ask how often planets get created *around* these stars. Is planet formation a very rare occurrence, or a common one? As we saw in our brief tour of history, this is an ancient question. And by the middle of the twentieth century, planet formation had once again become the subject of intense astronomical debate.

Drake expressed this question in terms of fractions. What, he asked, is the fraction of stars that host a planet? He wrote this term as the symbol f_p (read as "f sub p").

3. The Number of Planets in "The Goldilocks Zone"

It is not enough to just ask if a star hosts a planet. The planet's orbit around the star is also a key factor in thinking about life, intelligence, and civilizations. If a planet is very close to its star, then the temperature on its surface will be so high that life gets fried down to its atoms. If, on the other hand, a planet's orbit is very large, its surface will be perpetually frozen and in near darkness.

At the time of the Green Bank meeting, Otto Struve's former student Su-Shu Huang had just finished work that showed how each star is surrounded by a "habitable zone of orbits." Huang defined this zone as the band of orbits where liquid water can exist on a planet's surface.[51] Liquid water is thought to be a key factor in allowing life to form and thrive. The inner edge of Huang's habitable zone was the orbit where a planet's temperature was just cool enough to keep surface water from boiling. The outer edge was the orbit where the temperature was just high enough to keep water on a planet's surface from freezing.

Drake and his colleagues at the Green Bank meeting needed to know how many planets (for those stars that had planets) were in the habitable zone. In other words, how many planets were on orbits that left their surfaces neither too hot nor too cold. Thus, the third variable in Drake's equation would be the average number of planets in a star's habitable zone, which is also sometimes called the "Goldilocks zone." This term is expressed as n_p (read as "n sub p").

4. The Fraction of Planets Where Life Forms

While the first three terms in Drake's equation dealt purely with issues of physics and astronomy, the fourth brings chemistry and biology into the discussion. Given a star with a planet in an orbit that leaves it with liquid water on its surface, what are the odds that the simplest forms of life will appear? Once again, Drake expressed this question in terms of a fraction, which he called f_l (read as "f sub l").

It's worth noting that discussions about f_l hinge on the chemical pathways taking nonliving matter into a self-replicating state. The formation of life from nonlife is called *abiogenesis*. Experiments done by Harold Miller at the University of Chicago in the early 1950s had already provided compelling evidence that abiogenesis might not be difficult to obtain on a habitable-zone planet.[52]

5. The Fraction of Planets Where Intelligence Evolves

The fifth term moves us from the biochemistry of life's origin into the dynamics of its changing forms. Assuming life begins on a planet, how often would evolution carry that life forward to intelligence? Drake expressed the fraction of planets where intelligence evolves with a term called f_i (read as "f sub i").

6. The Fraction of Planets with a Technological Civilization

The sixth term moves us from evolutionary biology to sociology. Given that a planet hosts an intelligent species, how often does a technologically advanced civilization then arise? This question was represented by the term f_c (read as "f sub c"), the fraction of planets where a technological civilization begins.

For practical purposes, Drake saw "technologically advanced" as meaning a civilization with the capacity to broadcast radio signals.[53] So, while the Romans were certainly a civilization, from Drake's point of view, they don't count as a technologically advanced one.[54]

7. The Average Lifetime of a Technological Civilization

The final factor in Drake's equation is the most haunting: How long does a civilization like our own last? Can we expect another few centuries before our global society flares out, or are there many millennia of development ahead? Assuming that technological civilizations have occurred often enough for an average to be well defined, what is their average lifetime?

With this last term (written as *L*), Drake was asking the others at the meeting to consider alien sociology on a deeper level. Some discussion was devoted to the overconsumption of resources, but given the heightened fears of nuclear war in 1961, aggression was the focus of Drake's final variable.[55] Are most civilizations as aggressive and warlike as our own? Do they become more peaceful as they evolve? How long, on average, can they last without destroying themselves?

ONE EQUATION TO BIND THEM

Each of the seven terms Drake chose to set the agenda for his Green Bank meeting was a problem that, in principle, had a quantifiable answer. Each contained its own compelling mysteries, and each was a step on a ladder to that overarching question: Are we alone?

To be specific, though—and the whole point of the meeting was to be specific—Drake's overarching question was: How many radio signal producing technological civilizations other than our own reside in the Milky Way galaxy? In the language of Drake's agenda, what is the value of N?

With all his subproblems mapped out, Drake was finally in a position to put them together into a single equation. Here it is, written out in mathematical form:

$$N = N_* f_p n_p f_l f_i f_c L$$

In words, Drake's equation says the number of exo-civilizations from which we can get signals equals the number of stars forming each year (N_*), times the fraction of those stars with planets (f_p), times the number of planets where life can form (n_p), times the fraction of planets where life actually does form (f_l), times the fraction of those planets that evolve intelligence (f_i), times the fraction of those intelligences that go on to create technological civilizations (f_c), times the average lifespan of those civilizations (L).

Here, you can see why scientists like equations so much. An idea that takes a mouthful of words to express gets captured pretty cleanly in just one short line of symbols.

On the morning of November 1, 1961, with the participants at Green Bank meeting gathered around the conference table, Drake stood and wrote his new equation on the blackboard. Scrawled in chalk like a haiku, it was never intended to be anything more than a guide, an overview, an organizing principle.

It turned out to be much more.

A search of "Drake equation" on the Google Scholar search engine returns thousands of papers. A similar search of Amazon brings back scholarly books, science fiction novels, T-shirts, and even a tungsten carbide ring imprinted with the formula. Since the Drake equation was introduced, it has appeared in a stunning number of scientific conferences, magazine articles, and documentaries.

"It amazes me to this day," Drake wrote later, "to see [the equation] displayed prominently in most textbooks on astronomy, often in a big, important-looking box." With humility, Drake added, "I'm always surprised to find it viewed as one of the great icons of science because it didn't take any deep intellectual effort or insight on my part. But then as now it expressed a big idea in a form that . . . even a beginner could assimilate."[56]

In considering the importance of the Drake equation, you have to begin with what it is not. It is not a law of physics. Einstein's famous equation $E = mc^2$ expresses a fundamental truth about the behavior of the world. It is a statement of our understanding about how nature works on its own. The Drake equation, on the other hand, is really a statement of our *lack* of understanding. It tells us what we would need to know to get a specific answer to a specific question: How many exo-civilizations are out there?

Before Drake, the scientific consideration of exo-civilizations was unfocused. What existed was a mix of unconnected musings in scientific journals, books, and popular articles. There was no structure for building a coherent program of study, either through theory or observations. By breaking the big question into seven smaller questions, Drake crafted a useful way to think about the problem that also left scientists something they could work on. It gave them something to do.

Each of the terms in the equation could be explored on its own, using whatever means were available. Astronomers could work on the first three terms; biologists could think about the two that followed; sociologists and anthropologists could explore the last two. Of course, most of the work would be speculative. But at least it would be speculation with a focus and a scientific foundation.

With time and patience, advances were made from all sorts of directions. Computer studies of chemical reactions provided insights into abiogenesis. Evolutionary studies of life on Earth showed how cognitive patterns leading to intelligence first appeared. And while some terms, like the average age of a civilization, might never be known, others, like the fraction of stars with planets, were thought to be within grasp at the time of the Green Bank meeting. In addition, while even the closest stars were fifty trillion miles away, the planets in our solar system were relatively close. If we could find even one example of life—in its simplest form—on Mars or anywhere else in our solar system, that would tell us something powerful about the first biology term.

What Drake's equation gave astrobiology was a way to think about itself. In the process, it changed how we understood life, civilization, and ourselves.

Drake's equation also ensured the success of the Green Bank meeting. Beginning with the rate of star formation and marching all the way through to the average lifetime of technological civilizations, the nine participants did their best to make informed estimates of the different terms. History shows they were a hopeful group. They assigned values relatively close to one for all the fractions. Most telling, though, they reserved their pessimism for Drake's final factor: the average lifespan of technological civilizations.

The capacity for a civilization to short-circuit its own evolution through self-destruction vexed the meeting's participants. It would go on to become a bottleneck in all thinking about searches for extraterrestrial intelligence (SETI). The Green Bank participants believed, as Drake later wrote, that "the lifetimes of civilizations would either be very short—less than a thousand years—or extremely long—in excess of perhaps hundreds of millions of years."[57]

In the end, the group agreed that the final factor was what mattered most. The number of stars was so vast that the galaxy could absorb a lot of what Drake and his colleagues considered pessimism regarding the other terms of the equation. But the galaxy also needed to be populated *now*. There needed to be an overlap in time between our civilization and theirs so that there would be signals for us to receive. That meant the other civilizations needed to last at least millions of years, which seemed like a stretch for the Green Bank group.

Just before the meeting ended, Drake and his colleagues broke out their one remaining bottle of champagne (Calvin's call from the Nobel Committee had come late on the first night of the meeting). As they raised their glasses, Otto Struve offered a toast. "To the value of *L*," said Struve. "May it prove to be a very large number."[58]

THE DAY CLIMATE CHANGED

In 1965, little more than three years after Struve's toast, President Lyndon Johnson would raise the same issue of civilizations' longevity in the far more specific context of our own fate. Speaking before a joint session of Congress, he said, "[T]his generation has altered the composition of the atmosphere on a global scale through . . . a steady increase in carbon dioxide from the burning of fossil fuels."[59]

It's remarkable to note that, more than fifty years ago, an American president was already aware of, and acknowledging, human-created climate change. Johnson had been briefed on the dangers of CO_2 increases by the famous climate scientists Charles Keeling and Roger Revelle, among others. So, not only was President Johnson aware of the issue, but he was already concerned enough to raise it before Congress. That single sentence in his address gives the lie to the claims of so many climate-change deniers that global warming is some kind of recent hoax. Indeed, the scientific understanding of our effect on the Earth dates back more than a century. As President Johnson's speech demonstrates, even fifty years ago, that understanding was firm enough to gain notice at the highest levels of policy and politics.

But there is a difference between a community of scientists, at the vanguard of their fields, glimpsing human-driven climate change and the culture as a whole metabolizing the story. A single speech by a president can't create the kind of intimacy that is the hallmark of humanity's most powerful narratives about itself and its place in the world. That takes time and the play of events. The industrial revolution, for example, didn't arrive as soon as the first factory was built. It took people moving en masse from farms to cities where day-to-day life took on new rhythms and textures. Only then did we begin to see ourselves as "industrial." Only then could we tell new stories about ourselves as a civilization that conquered the planet with steel, rubber, and oil.

Likewise, we are just beginning our entry into the Anthropocene. Fifty years on from President Johnson's speech, we have just started becoming familiar with images of melting glaciers, massive heat waves, and flooded cities. We are just beginning to experience what life on a climate-changed world looks like. But when President Johnson stood before Congress in 1965, that story was still new.

Conservation was the intended theme of the president's speech that day. Only a few years had passed since biologist Rachel Carson had raised alarms over the environmental effects of pesticides in *Silent Spring*. Even less time had elapsed since the treaty banning atmospheric testing of nuclear weapons had gone into effect. While the Cold War made instant annihilation a credible threat in the 1950s, by the mid-1960s some were beginning to realize that even the everyday activities of our project of civilization were not, in total, going unnoticed by the planet.

Sustainability on a global scale, however, is a very different kind of story for humanity to tell itself. It demands a vastly enlarged imaginative palette. At the time of President Johnson's speech, the picture of a climate-threatened future was just starting to be painted by scientists as they gained a first foothold on understanding the Earth as a planet. These researchers were recognizing, for the first time, that Earth needed to be understood in its entirety as single, tightly coupled system—a kind of vast, planetary-scale machine.

Ironically, and as is so often the case, the need for this new vision found its first urgency in the needs of warfare. With the rise of long-range bombers and intercontinental missiles, cold warriors were busy imagining Earth from well above the atmosphere. But they were also deeply concerned with how weather could tip the scales of battle. It was partly at their urging that resources poured into the scientific study of climate. A nuclear-powered laboratory was built under the ice of Greenland to understand how weather patterns changed over the course of millennia. Instrument-laden ships crisscrossed the oceans, studying the forces driving deep ocean currents. Most importantly, the same ICBMs threatening nuclear war were starting to lift scientific satellites into orbit, where their eyes would point downward to study the Earth.

These were expensive and globally extensive efforts. They laid the groundwork for a new vision of our project of civilization's planetary context and impact.

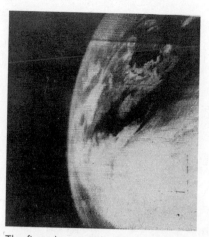

The first photograph of Earth captured by a weather satellite, taken in 1960.

In 1960, a still-wet-behind-the-ears NASA launched its first successful weather satellite, TIROS (Television Infrared Observation Satellite). By 1962, TIROS was offering continuous coverage of the Earth's weather.[60] In the wake of TIROS, no longer would a hurricane unleash its violence on an unsuspecting population. And for the first time, people were treated to images of Earth as a globe suspended in space. Even the earliest grainy videos showed the elegant arc of the world's horizon as seen from high above the atmosphere, a vision that would rewire our collective imaginations.

By the mid-1960s, a convergence had begun. Images from TIROS, President Johnson's address on carbon dioxide, Fermi's lunchtime insight, and Drake's Green Bank conference were pieces in a cultural jigsaw puzzle that was beginning to assemble itself. Each represented a tentative first step toward seeing our project of civilization in a new light—the light of the stars. Fermi and Drake represented a new awareness among scientists that the story of our own project of civilization must be set onto a cosmic stage, with all its stars, planets, and possibilities. Meanwhile, studies of climate funded via Cold War urgencies shaped an awakening among other scientists that Earth's story must be told in terms of a mighty planetary system driven by sunlight and shaped by life—including our own. Finally, President Johnson's address signaled that our civilization's impact on the planet was making its way into the domains of culture and politics.

A new human story, a new human mythology, was emerging. The outlines of this narrative, in which human beings and our project would be inescapably bound to the machinery of planetary evolution, were beginning to take shape. Few at the time could recognize the power, the peril, and the promise growing in this new story. It was still too new and too unformed. To take the next steps in forging this new vision, we would have to leave home. We would have to become wayfarers and journey, for the first time in our long history, out to the high frontier of space. That was where the sibling worlds of our solar system were waiting to tell us their secrets.

CHAPTER 2

WHAT THE ROBOT AMBASSADORS SAY

TO BE A BUM

The Florida sun glistened over the blue Atlantic waters, but Jack James was in a black mood. It was July 22, 1962, and it had been a very bad day. The Texas-born engineer was project manager for NASA's Mariner program, which aimed to send America's first emissary to another planet. James had gotten the job a little more than two years earlier. Like everything else in the space race of the early 1960s, James's program had been rushed forward at breakneck speed, working nonstop. Now the fruit of all that effort lay in ruins at the bottom of the ocean.

James and his team had been given less than fourteen months to design, build, and launch a probe to Venus.[1] Until then, the Moon had been the primary target of the space race, and America's record for hurling oddly shaped boxes of electronics at Earth's rocky satellite had been a mixed bag. The Russians were having better luck. They'd gotten three of their nine probes to the Moon. The US had only gotten one there.[2] Now NASA was desperate for more than a

Rocket engineer Jack James (right), with Mariner
Project manager Dan Schneiderman (left).

win. It needed to upstage the Soviets in a big way. That's why James
was given the audacious job of thinking beyond the Moon and going
interplanetary.

Mariner 1 was designed to perform a "fly-by" past Venus, a
world that orbited 30 percent closer to the Sun, but with almost
the same mass and size as Earth.[3] The probe's original design called
for 1,250 pounds of scientific instruments, communications gear,
solar panels, rocket motors, and fuel. But the new, more powerful
generation of rocket boosters Mariner was supposed to ride into
space kept blowing up, and NASA brass soon demanded a redesign.
James's engineers were forced to quickly shed more than two-thirds
of Mariner's weight.

James navigated his team through every design change and
every challenge. That was how they came to this day. With the whole

world watching, *Mariner 1*'s booster rocket lit up the Florida sky as it blasted off that July morning. For the first few moments, the launch looked clean. But then the Atlas booster began to fishtail. Every launch has a "range safety officer" whose job is to blow the rocket to bits if it looks as though the mission is failing and it's going to crash back to Earth. Four minutes and fifty-three seconds into the flight—and just six seconds before *Mariner 1* would have safely separated from the main launch vehicle—the safety officer hit the big red button.[4]

Boom!

For a full minute after the rocket exploded, telemetry continued to get signals from the probe as it tumbled from the sky to its ocean grave.[5] At least *Mariner 1* had been a tough bird.

They'd been so damn close. Just six seconds more and they'd have been on their way to Venus.

"Born to Lose," by Ray Charles, played on the radio as James drove back to his rented apartment in Cocoa Beach. He was in despair. Years later, he'd recall the mantra of all space engineers: "To be a hero there are ten thousand parts that need to work properly on a spacecraft. To become a bum you just need one of them to fail."[6]

But while James and his team were down, they weren't out. The now-destroyed probe had a twin. *Mariner 2* was waiting back at Cape Canaveral.[7] There was still time to be a hero.

THE VENUS PROBLEM

The logic of the space race dictated that either Mars or Venus would be the next destination after getting probes to the Moon. Both were neighbor planets that could be reached in a matter of months, a step up from the three-day trip to the Moon. And each had its own long history of dreamers imagining a temperate world fit for extraterrestrial life.

Given its proximity to the Sun, Venus gets twice as much solar energy as Earth.[8] That's why many early astronomers imagined Venus as a jungle planet. In 1870, Claude Flammarion (the author of *The Plurality of Worlds*) thrilled his readers with images of a Venusian landscape made of broad, swampy plains ringed by mountains higher than the Himalayas.[9]

Flammarion assured his readers that Venus was a world rich

An imagined view of Venus from Flammarion's 1884 book, *Les Terres Ciel*.

with life: "Of what nature are the inhabitants of Venus . . . ? All we can say is that the organized life [there] must be little different from terrestrial life, and that this world is one of those that resembles our own most."[10]

But with the increasing power of astrophysical observations, this pleasant dream of a jungle Venus would come under fire. First, astronomical observations in the late eighteenth century revealed Venus to be perpetually shrouded by clouds.[11] Then, in the mid-twentieth century, the Venusian atmosphere was revealed to be heavy with carbon dioxide (CO_2). The Earth's atmosphere is 78 percent nitrogen, 21 percent oxygen, and one percent everything else. CO_2 comes in at a mere 0.039 percent of the air you are breathing right now. That's a pretty small fraction for a molecule that, as we will see, has a big role to play in our story. But for Venus, CO_2 is pretty much all there is to the atmosphere, accounting for more than 95 percent of all its gases.[12]

The presence of so much CO_2 was bound to make Venus a very different place from Earth, and by 1956, astronomers had gained their first evidence of just how different it might be. Using the same kind of radio astronomy technologies Frank Drake would soon employ in Green Bank, scientists from the Naval Research Laboratory found evidence that Venus's surface temperature was well above 600 degrees Fahrenheit.[13] That's hundreds of degrees above the boiling point of water. If the NRL result was true, then Flammarion couldn't have been more deluded. His Venusian swamps would have boiled away long ago. More importantly, 600 degrees was far too hot for any form of life to survive. It seemed the place Venus resembled most wasn't Earth, but Hell.

While geology and its study of the Earth had been around a

long time, planetary science—which takes all planets as its subject—
was a young field. The NRL results set off a firestorm among the
small group of researchers who considered themselves planetary
scientists. Part of the conflict came because, just a few years earlier,
another team had predicted Venus to be covered by a vast, planet-
girdling ocean.[14] But neither oceans nor lakes nor even cups of tea
could be squared with the new NRL data, which suggested temperatures on Venus comparable to the inside of a pizza oven.

In response, some scientists claimed the NRL's data had been
misinterpreted. Its source, they claimed, wasn't Venus's surface but
violent, atomic-scale processes occurring at the boundary of its
atmosphere and the harsh conditions of interplanetary space.[15]

Resolving the dilemma required more power than earthbound
instruments could provide. The best telescopes of the day could not
see the disk of Venus in enough detail to distinguish between a hot
surface or processes occurring high in the atmosphere. Getting up
close with a space probe was one means of getting that level of detail.

But a space mission wasn't the only key that astronomers
needed to unlock the mystery of Venusian climate. The NRL result
was shocking to scientists because no one could understand how it
might be true. Venus is closer to the Sun, but that proximity should
only raise its surface temperature a few tens of degrees, not hundreds.[16] If the surface temperatures really were 600 degrees, how
could a planet so like the Earth in so many ways have ended up so
different from our world? What was needed was a theory explaining how Venus might end up with such insanely high temperatures.

That task would be taken on by the young, untested, and not-yet-
minted PhD student Carl Sagan. Though no one at the time could

have guessed it, not only would Sagan's work solve the Venus problem, it would also set the stage for a deeper understanding of our own world's entry into the Anthropocene.

THE GREENHOUSE EFFECT

Though he died in 1996, Carl Sagan remains one of the most recognizable scientific faces in the popular imagination. Born sixty-two years earlier to working-class Jewish parents in Brooklyn, Sagan's love affair with science began as a young boy during a trip to the 1939 World's Fair. The passionate interest in life on other planets that defined his life came a bit later, as a teenager, with a steady diet of *Astounding Science Fiction* magazine and writers like H.G. Wells.[17]

Sagan attended the University of Chicago, where he was trained to think as both a scientist and a humanist. It was a combination that would later prove so compelling to millions via his popular writings and television programs. After Chicago, he moved ninety miles northwest to the Yerkes Observatory in Wisconsin, where he started work on his PhD.[18]

Graduate work in astrophysical sciences takes years of dedicated effort. First, there are advanced classes on the basics of theory and observation. Only after this initial phase can students start independent study. Sagan arrived at Yerkes with an interest in life beyond Earth, so for his graduate thesis he chose three separate issues at the intersection of planetary science and what we now call astrobiology. The first of these would be the Venus problem.

Sagan's question was straightforward: What process could turn the surface of Venus into a scalding hell? Combing through decades of scientific literature in search of an answer, he found one in what is now well known as the greenhouse effect.

A planet like the Earth would be a deep-freeze world without its atmosphere. That conclusion requires only a few lines of basic physics to demonstrate. Sunlight hitting a planet warms its surface. The warmed ground emits what is called heat radiation, which is just electromagnetic waves generated by the jiggling motions of heated atoms. Any object at any temperature above absolute zero spews heat radiation into its surroundings. That includes your own body as you read these words.

For our planet's temperature to remain steady and unchanging, the energy flowing onto it must balance the energy flowing out. Heat is just another form of energy. That means incoming solar energy and outgoing heat radiation energy must balance if the Earth's temperature is to stay constant. Scientists call this balance the planet's equilibrium temperature.

Calculating a planet's equilibrium temperature requires the kind of basic physics most students learn in their first-year astronomy classes. Once they work through the math, those students all come up with the same startling result: Earth without an atmosphere would have an equilibrium temperature around zero degrees Fahrenheit. That's well below the freezing point of water.[19]

As we all know from daily experience, most of Earth's surface is not frozen. In fact, the planet's current average temperature is a balmy 61 degrees Fahrenheit.[20] Somehow, our planet manages to stay warm enough for most of its water to be in liquid form, rather than as a solid (ice) or a gas (water vapor). It's the atmosphere that raises the temperature. The blanket of gases surrounding the planet keeps Earth's equilibrium temperature well above freezing. But how, exactly, does that happen?

The fact that you can see the Sun on a cloudless day gives testimony to the fact that Earth's atmospheric gases are mostly transparent to our star's incoming *visible* radiation. The Sun's visible-range electromagnetic waves pass right through our atmosphere as unmolested as through a clean glass window. But the heat radiation emitted by the warmed planet's surface isn't in the visible part of the spectrum. Instead, the planet radiates at longer *infrared* wavelengths the eye can't see. So, while incoming sunlight passes freely through the atmosphere, for the longer infrared wavelengths emitted by Earth's warmed surface, it's a different story entirely.[21]

Like a blanket you throw over yourself on a cold winter night, the blanket of gases surrounding our planet holds in energy that would otherwise get radiated away. It's this trapped energy that raises Earth's temperature above freezing. An actual greenhouse works along a similar principle, as the windows allow sunlight in but keep warmed air from rising away—hence the name, the greenhouse effect.

The greenhouse effect was old news for scientists studying the Earth. In 1896, Swedish Nobel Prize–winning chemist Svante Arrhenius had discovered the human impact on Earth's greenhouse effect.[22] Using a simple mathematical model, Arrhenius laid out the physics of Earth's greenhouse warming, demonstrating how our planet was warmed by its atmosphere. Just as important, his calculation also revealed how our own activity was adding to that warming. Using records of coal consumption, Arrhenius saw we were already putting enough CO_2 into the atmosphere to change the energy balance. Using the coal data, he predicted that human beings would eventually raise the planet's temperature as we continued dumping CO_2 into the air. His pencil-and-paper calculation predicted a global increase of about five degrees.[23] This is remarkably close to modern estimates. In our current era of climate-change denial, it's startling to recognize how far back the understanding of human-driven climate change begins.

Sagan wanted to go farther than Arrhenius—literally. He saw that what was true for the Earth must also be true for distant planets. The greenhouse effect had to be universal. So Sagan set himself the task of calculating the extent of the greenhouse effect on Venus to see if it could explain that planet's extreme temperatures. Across many cold Wisconsin winter days, Sagan pored over old papers in the Yerkes library, teaching himself the basic physics of infrared atmospheric absorption and its subsequent planetary warming. After months of exhausting work, he had his answer. With its CO_2-rich atmosphere, Venus was trapping enough energy to raise the surface temperature near to the staggering 600-degree level implied by NRL data.[24] The planet was a cauldron because of the greenhouse effect.

Today, scientists recognize that planets anywhere in the universe must be subject to the same set of forces and processes. While each world has its own unique story, those stories are all enacted by the same list of players: the flow of winds, the pull of gravity, the dance of chemistry. Earth is no different, and this, as we will see, is the principal lesson of the Anthropocene. But when Carl Sagan was working alone in the Yerkes library, the application of this universal vision of the universe's planets was still young. Other than a few nearly forgotten studies, Sagan was alone in bringing the earthbound process of greenhouse warming to another world. "Almost nobody on the planet as far as I could find, was interested in the

Venus greenhouse effect . . . ," he would later recall. "I sort of stumbled on it myself."[25]

A LIVING HELL

Rocket engineer and project manager Jack James only had a day to mourn the loss of *Mariner 1*. The launch window in which Earth and Venus were positioned just right for the calculated flight path would close in a month. James's team needed to get *Mariner 2* ready for launch immediately. Twenty-eight days later, at 2:53 a.m. on August 27, 1962, another Atlas-Agena rocket lifted from the ground atop another pillar of fire.

This time, the launch was successful, but just barely. A few seconds before the Atlas booster was to separate from *Mariner*, one of the rocket's control engines shut down, driving it into an uncontrolled spin. As fears rose for another failure, the first of the mission's "seven miracles" occurred. Control was regained at just the right moment to undo any damage the spin had imparted to the probe's calculated flight path. The rocket's second stage fired and *Mariner 2* was on its way to Venus.

It would take three months for the probe to cross more than twenty-five million miles of interplanetary space. Six more times, critical elements in *Mariner*'s systems would fail: a solar panel stopped working; temperatures on the space probe climbed to dangerous levels; the onboard computer failed to switch instruments to "encounter mode" as Venus approached. But each time, disaster was averted as the problem either fixed itself or James's team from NASA's Jet Propulsion Laboratory (JPL) found a workaround.[26]

"I'd get called at all times of the night," James recalled later. "My nerves had become so taut by this time that I instructed everyone that called me to start out with one of two sentences: 'There is no problem,' or, 'There is a problem.'" More than a few calls began with "There's a serious problem."[27]

In spite of all the difficulties, on December 14, 1962, *Mariner 2* flew within twenty-two thousand miles of Venus, a distance about six times the diameter the planet. As data from *Mariner 2* trickled into JPL, it became clear that the NRL study and Carl Sagan's greenhouse effect theory had been right. The scalding temperatures were not high in the atmosphere, but down on the planet's surface. Venus was indeed a living hell.[28]

The evidence for Sagan's greenhouse model for Venus got stronger as the space age matured. Over the next forty years, more than twenty other probes would visit our sister planet. Some mapped its surface at high resolution via cloud-penetrating radar. Others made detailed explorations of atmospheric conditions, including winds whipping around the planet at hundreds of miles per hour. The Russians even managed to get probes down to the surface. The probes worked for just a few hours before succumbing to the planet's intense heat and nuclear submarine–crushing pressures.[29]

What emerged from these studies was a picture of a world where the CO_2 greenhouse effect had run amok. The catastrophe was called a runaway greenhouse effect, and its discovery proved to be essential for understanding the climate cycles that run our own world.

The principal way that CO_2 gets added naturally to a planet's atmosphere is through volcanic eruptions. Molten rock explodes through the surface, venting huge amounts of CO_2. Radar imaging of Venus shows ample evidence for volcanism in the recent past (meaning the last hundreds of millions of years). But what volcanoes give, water can take away. "Weathering" by water, in the form of rain and rivers, breaks rocks down to their chemical components. Later, these molecular components can bind with CO_2 and get packed back into solid forms—that is, as new rocks. This is the basic process that creates what are called "carbonate" minerals like the limestone under Miami.

So, CO_2 belched into a planet's atmosphere via volcanoes can go back into the ground in rocks. Eventually, the rocks are subducted (dragged down) into lower regions of the Earth, where they melt, allowing the CO_2 to find its way back into the atmosphere through future volcanoes. It's a cycle that regulates the carbon dioxide in the atmosphere, and therefore the planet's greenhouse effect. It's also a cycle that appears to have been broken on Venus.[30]

At some point, Venus likely had more water. It may even have had oceans and been hospitable to life. But when some of that water evaporated, it made its way high into the atmosphere, where a deadly process began. Close to the edge of space, ultraviolet radiation from the Sun (the same kind that causes skin cancer) zapped the water molecules and broke them apart into hydrogen and oxygen. Hydrogen, being the lightest of all elements, easily escaped into interplanetary space as soon as the water molecules were broken apart. With the hydrogen gone, there was no chance for the broken water molecules to reform. Over time, and high in its atmosphere, Venus was bleeding its precious water into space.[31]

The planet's water loss resulted in what scientists call a positive feedback loop on climate. More water loss meant less rock erosion and less CO_2 bound up in rocks. More CO_2 in the atmosphere meant a more pronounced greenhouse effect and higher temperatures. But higher temperatures meant more water loss, which . . . well, you get the picture.

On Earth, there is no danger of losing our water in the way that Venus did. Our planet's atmosphere has a particularly cold layer, about twelve miles above the ground, that causes water to condense out and fall as rain or snow, keeping it from ever making it to very high altitudes. This "cold trap," as scientists call it, may have existed at one time on Venus. But at some point, its atmospheric layers changed, allowing water molecules to begin diffusing up to the heights where they could be split apart and lost forever.[32]

With its water safely trapped closer to the surface, Earth's carbon cycle acts as a negative feedback on climate. Negative feedback cycles keep small changes in temperature from growing out of control. Imagine if Earth's temperature were to jump by a few degrees. The negative feedback begins when this higher temperature leads to more evaporation. Then more evaporation leads to more rain; more rain leads to more weathering; and more weathering leads to more atmospheric CO_2 being drawn into rocks. Now there's less CO_2 in the air, meaning the greenhouse effect is reduced and the planet's temperature comes back down.

By giving us an explicit example of the greenhouse effect gone wrong, Venus helped teach us about the effects of negative and positive feedback loops on planetary climate. It made us think more deeply about the cycles of matter and energy that give a planet its character—or cause that character to change. From *Mariner 2* onward, the probes we sent to Venus let us see exactly how a planet that might have been a kindred twin had, instead, become a monster. Using the early understanding developed purely from studying Earth, the Venus missions allowed us to flex the muscles of a young climate science and broaden the reach of its knowledge. Like a doctor studying pathological cases of a disease to understand the basic workings of a healthy physiology, Venus's runaway greenhouse became a laboratory for understanding the complex interplay of atmospheres and geology that shape a world like ours.

By taking our first steps toward the planets, we were also taking the first steps toward understanding the laws of planets. We were beginning the process of using the worlds of our solar system to unpack the general and generic laws all planets must obey. Our early missions to the planets, led by pioneers like Jack James and early theoretical studies by Carl Sagan, were also our first steps in growing up as a planetary species. We were seeing, for the first time, the depth of our commonality with the rest of creation.

It's worth noting that, while Carl Sagan got the credit he deserved for predicting Venus's hyperactive greenhouse effect, his name was not on the paper reporting *Mariner 2*'s results.[33] Early in the project, Sagan had been put on *Mariner*'s design team, where he had, among other things, argued for a camera to be included on board (his proposal was rejected). But as Jack James's group pushed hard to make its deadlines, some felt Sagan was not pulling his weight. Their misgivings proved to be correct. A crisis had erupted in Sagan's personal life that kept him from making the expected contributions to the mission.

In 1957, when he was still working on his PhD, Carl Sagan married Lynn Margulis, a brilliant but as-yet-undirected student (at the time, her last name was Alexander). When they met, Margulis had not yet settled on science as a career. Sagan helped introduce her to questions concerning life and planets. A fire was lit in the young woman's imagination, and even as their children were born, she took on the task of graduate work in biology. But Sagan's relentless work schedule left the full burden of raising their children and managing the household to Margulis. After five years of trying to hold the demands of family and graduate work together, Margulis had had enough. She packed up the children and left Sagan to his overcommitted work schedule.[34] But in one of the great turns of scientific history, Lynn Margulis would return to play an equally important role in understanding the coupled histories of life and planets. Before that story could play out, however, Sagan and the rest of the world would have to come to terms with Mars.

BEDROCK MARS

Steven Squyres, the chief scientist for the multibillion-dollar Mars Exploration Rover program, was not nervous. Sure, the plan was insane, but that didn't mean he had to be nervous. It was January 25, 2004, landing night for the robot rover *Opportunity*. Squyres was waiting in the flight control room at NASA's Jet Propulsion Laboratory while, more than three hundred million miles away, the *Opportunity* rover was bundled in its descent capsule, hurtling at twelve thousand miles per hour toward Mars. Since blasting off from Earth six months earlier, *Opportunity* had been on a direct path toward the Red Planet. But it wasn't going to slow down and ease into orbit, as in some previous missions. Instead, the $400 million probe was on a straight shot toward its landing zone in the Meridiani Planum, a broad plain just south of Mars's equator.

The entry, descent, and landing (EDL) phase called for *Opportu-*

nity to dive straight in from space, shedding speed via friction with Mars's thin atmosphere. A supersonic parachute would then blow open, slowing the capsule further. After that, if all went according to plan, the lander would spool down, away from the rest of the spacecraft, via a sixty-five-foot-long tether. As the descent continued, a cocoon of giant airbags would explosively inflate around the lander. Approximately one hundred feet from the ground, retro-rockets would fire, bringing the whole spacecraft to a halt. The lander, surrounded by its airbags, would hang forty feet from the ground. Then the tether would be cut away, dropping the airbag-enshrouded lander to the surface, where it would bounce like a beach ball on steroids. Eventually, after a mile or so of bouncing, the lander was supposed to come to a safe resting place on the Martian surface.[35]

Yeah, the idea was insane.

But it was an insane idea that had already worked. Just three weeks earlier, *Opportunity*'s twin, the *Spirit* rover, had bounced to safety on the other side of the planet. That six-wheeled mobile geology laboratory was already wandering the Martian surface, taking data. So Squyres was not nervous. Well, not too nervous.

There was a long wait as the JPL flight team searched for signals that *Opportunity* had survived its ordeal. Then the EDL manager yelled out to the room, "We're down, baby!" The room exploded in cheers. *Opportunity* was safe on the surface.

Within the hour, Squyres switched to the rover operations room as *Opportunity*'s cameras came on and his team tried to see exactly where their creation had come to rest. "The picture comes [up on the screen] and it's dark," Squyres recalled later. "There's something there but it's underexposed." Slowly, the image gets calibrated, or "stretched." "The stretch hits and instantly I realize what I'm seeing," Squyres writes. "It's impossible, it's too good to be true, it's too good to believe."[36]

Right in front of the rover was an exposed layer of bedrock—the kind of thing you see on Earth when you're driving on a road cut through hills. And, just as on Earth, the layer of exposed rock Squyres was staring at represented a record. It was a sandwich of compressed Martian history going back millions or billions of years. They were staring at Mars's planetary evolution written in rock: the scientific equivalent of pure gold.

THE RED PLANET SHUFFLE

In a world of instant electronic access to all human knowledge and of routine jet travel five miles above the Earth, it's easy to miss the audacity of the Mars rovers. Getting *Spirit* and *Opportunity* (and, later, *Curiosity*) safely on Martian ground was crazy enough. But the genius embodied in the rovers is reason to be proud of humankind. These robot scientists have trundled across miles of Martian landscape, drilling into rocks, sniffing for critical chemical compounds, and imaging the Red Planet at high resolution. The missions represent the best of our collective vision and capacity for solving the most challenging problems.

But the exploration of Mars by these rovers and other international probes represents something else that transcends engineering. Each was a step on the ladder of our coming of age as a planetary species. By literally giving us visions of another world through their high-resolution cameras, a new understanding of other worlds—

and, perhaps, other worlds with civilizations—could be born. But climbing to that understanding was fraught with difficulties, as reality shattered our expectations and then shattered them again.

Like Venus, Mars was an early target of our interplanetary explorations. Just two years after Jack James's JPL team flew *Mariner 2* inward toward the Sun, their *Mariner 4* probe made the journey outward to Mars, a planet with an even longer and more storied place in our extraterrestrial imaginings.

For the *Mariner 4*, mission Carl Sagan was again on the design team, and this time he won the debate about cameras. *Mariner 4* carried a primitive (by today's standards) analog TV camera. The pictures it sent back instantly changed our dreams of what Mars might be and what it might mean to us.

Because of Venus's eternal cloud cover, it never appeared as anything more than a white disk. But for Mars, the story was very different. By the mid-1800s, astronomers knew Mars had surface features that changed over time. This led many nineteenth-century scientists to a dramatic conclusion: Mars had a climate like our own.[37]

Most importantly, astronomers saw that Mars had that most essential of climatic features: seasons. White polar caps on the Red Planet had been seen as far back as the seventeenth century. The polar caps grew and retreated as Mars progressed through its 687-day orbit. It was with good reason that in 1870, Claude Flammarion envisioned Mars as a world rife with life.[38]

By the turn of the twentieth century, the Mars story gained a new level of drama via Percival Lowell's obsession with the Red Planet. Lowell's fascination had begun with earlier studies by the Italian astronomer Giovanni Schiaparelli, which appeared to show long, straight features on the surface. Lowell claimed these were canals, representing the work of an intelligent civilization.[39] In popular books, Lowell argued forcefully that Mars was inhabited and its society was, in effect, a victim of climate change. The planet was drying up and the canals were a desperate attempt to bring water from the polar ice caps. While most astronomers dismissed Lowell's observations as wishful thinking, in the popular imagination the die had been cast. Through books like H.G. Wells's *War of the Worlds*, Mars became the alien world most people imagined to host an alien civilization.

By the mid-twentieth century, astronomers had already accumulated enough telescopic evidence to be confident that Mars was not home to an advanced civilization. The atmosphere appeared to be thin and the planet cold. Still, the possibility that life existed in some form on that world remained very real. Periodically, the planet experienced significant changes in color that some argued had a biological origin.[40] As *Mariner 4* was launched, Carl Sagan remained hopeful that Mars might be home to at least some kinds of vegetation or, at the least, microbes.

But when *Mariner 4* sailed past the Red Planet on July 14, 1965, the twenty-two images it sent back killed the dream of life on Mars in both the public and scientific imaginations.

It was the craters that did it.

Mariner 4 saw *a lot* of craters on Mars, and some of them were vast. On Earth, craters don't last long. Whether they form from volcanoes or from meteor impacts, most craters on Earth get erased after many millions of years. It's the familiar processes of weathering by wind and water that wipe the craters away. Seeing large craters on Mars meant its surface hadn't changed in billions of years.

Mariner 4 showed us a Mars that looked a whole lot like the empty, desiccated Moon.[41]

In the wake of the new pictures, a *New York Times* editorial announced to its readers, "The astronomers of past decades who thought they detected canals on the Martian surface and speculated that it might have bustling cities and beings engaged in lively commerce were victims of their own fantasies." Concluding, "The red planet is not only a planet without life now but probably always has been."[42]

First Venus and then Mars. The main accomplishment of humanity's first interplanetary emissaries seemed to be the death of our interplanetary dreams of life on other worlds.

Luckily, Mars didn't stay dead for long. In 1971, *Mariner 9* became the first spacecraft to park at a planet's doorstep. Rather than just zipping by at ten thousand miles per hour, *Mariner 9* went into orbit around the Red Planet. By taking up residency this way, the probe found Mars's story to be far more complicated and far more interesting.[43]

Mariner 9 was built to map a good deal of the planet's surface. When it arrived, however, it found Mars covered in a globe-swaddling dust storm. The surface was totally obscured. Because the space probe had been built with some inherent software flexibility, NASA engineers were able to delay the mapping till the storm abated (two Russian probes that arrived at the same time as *Mariner 9* had no such flexibility and returned little useful data). While *Mariner*'s work was delayed, the planet-encircling storm highlighted the critical role airborne particles (that is, dust) could play in shaping climate.[44] In the years to come, that link would become a political football for earthbound policy makers.

Eventually, the storm cleared and *Mariner 9* returned more than seven thousand images. In those pictures was our first hint that, while today's Mars may be bone-dry and frozen, Mars of the past might have been a very different kind of world. The pivot depended entirely on water.

Mariner 9 revealed landscapes that looked a whole lot like they'd been carved by flowing water. There were dry riverbeds and broad deltas. There were floodplains and rainfall basins. Confirmation that these features really were shaped by torrents of liquid water would have to wait for future missions. But what *Mariner 9* immediately told us was simple and profound: the planet had changed in a big way.[45]

Mariner also revealed that our smaller neighbor was a planet as unique as our own. Mars was home to Olympus Mons, a towering

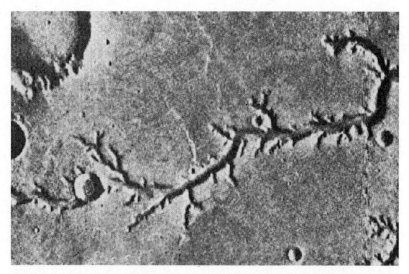

The view of the Nirgal Vallis channels on Mars taken by *Mariner 9* in 1971. Images like these were the first indication that Mars once had water flowing on its surface.

volcano that rises almost fourteen miles from the planet's surface. It also hosts Valles Marineris, a four-mile-deep canyon the size of North America that put the puny crack in Arizona we call "Grand" into a new, cosmic perspective.[46] Mars, it turned out, had volcanoes and valleys, craggy highlands and smooth, broad lowlands. It was a place all its own, with tourism-worthy sites unlike anywhere on Earth. And all this topography would matter as the first attempts to understand the Martian climate got underway.

The next great step in reviving the possibility of life on Mars came with the two *Viking* landers that touched down via parachutes and retro-rockets in the summer of 1976. Once again, Carl Sagan played an integral role, designing lander experiments that looked for microbial life in the Martian soil. The biology experiments returned ambiguous results, but the *Viking* landers' meteorological stations allowed us to see, for the first time, what the weather was like on another planet.[47] Each Martian day (called a *sol*), the *Viking* landers sent back measurements of temperature, pressure, and wind. The data flowed for six years, until one lander failed and the other was turned off by mistake.[48] Through *Viking* we were on our way toward seeing weather and climate on other worlds as a cousin to our own.

With the advent of the Martian rovers in the 2000s, the mantra of NASA's Mars program became "follow the water." If life had once existed in Mars, we'd first have to prove the planet was once wet enough and warm enough to support life.[49] But the presence of surface water can never be separated from the question of climate. So by following the water, NASA also committed itself to unpacking the story of Martian climate *and* Martian climate change. Like

Venus, the Red Planet was acting as a guide for understanding our own world.

THE GREAT MARTIAN CLIMATE MACHINE

Robert Haberle wasn't planning on becoming a world expert on the Martian climate. After serving in Vietnam, Haberle returned to civilian life in 1968, kicking around Europe for a while, "being young and anxious to explore the world." Finally, starting in college at San Jose State, he needed to declare a major. "I was looking through the catalogue and saw meteorology," he recalled to me in an interview. "I thought that meant the study of meteors. My wife had to explain to me it was about the weather."[50] It was an unlikely beginning for a man who would eventually help develop NASA's premier Mars Global Climate Model, one of the world's most powerful tools for studying the Red Planet's history.

The model's own history dates back to the late 1960s, when pioneers Conway Leovy and Jim Pollack took a climate model developed for Earth and began adapting it for Mars.[51] Pollack was one of Carl Sagan's first graduate students, and they collaborated together for years. Leovy was an atmospheric pluralist. He wanted to build a version of climate studies that reached beyond Earth to embrace every planet with an atmosphere.

For scientists, the word *climate* refers to long-term patterns of weather. While the weather changes from day to day (sunny on Tuesday but raining on Wednesday), climate represents the long-term patterns of winds, precipitation, ice cover, and ocean flow. To make a climate model, scientists must solve the physics equations governing these processes. That means a climate "model" is really a mathematical physics model. It's a description of the world that uses the highly specific and very exacting language of mathematical physics.

Just as architects make models of a skyscraper out of paper, balsa wood, and plastic, scientists use the laws of physics, expressed in the language of mathematics, to construct models of a physical system. If it's a gas engine they're modeling, then the mathematics lets them understand and predict something like the engine's fuel consumption. If it's a bridge they're modeling, then the mathematics lets them understand and predict how many cars can safely travel from one side of the bridge to the other. And if it's a planet's climate they're modeling, then the mathematics lets them understand and predict the long-term patterns of temperature, cloud cover, and so on.

To be effective, however, a climate model needs a lot of "moving parts." It needs to describe a lot of different kinds of physics, chemistry, and, perhaps, other processes as well. It must account for the flow of atmosphere on a spinning planet. It has to describe how radiation from the Sun warms the air near the surface, causing gases to rise. It must deal with how some of those gases, like water vapor or carbon dioxide, will condense into liquids or ice when they get cold (that's how the models track cloud formation, rain, and snowfall). Building a climate model that gets the answers right (meaning it matches observations) requires years of insanely hard work.

It also requires a lot of equations to describe the combined action of atmospheric flow, condensation, and the movement of radiation. Each one is pretty complicated on its own, taking a lot of human ingenuity to master. But solving all the complicated equations together at the same time is simply beyond the intellectual power of any one person. So to make progress, scientists must turn to digital computers that solve the equations in tiny steps, over and over again, billions of times each second. In this way, the computers *animate* the equations. They bring details hidden in the mathematical complexity to life. And the models Haberle and others built did just that. They brought Martian climate to life for scientists. Through the models, researchers could see the full complexity of Mars's climate. Most important, they could see both the similarities and the differences in how it worked relative to that of our own world.

JUST LIKE EARTH, ONLY IT'S NOT

"All planets are subject to the same basic forces," says Robert Haberle. "It's just that the strength of those forces will be different on different planets."[52] While Mars today may be a frozen, arid world utterly unlike Earth, the mechanics of its climate bear essential similarities to ours. Let's start with its differences from Earth. While Venus has a lot more atmosphere than our planet, Mars has a lot less. The surface pressure read off by the *Viking* landers and the other Martian weather stations is less than one percent of what we get on Earth. That means the total weight of Mars's blanket of gases is 99 percent less than Earth's. Like Venus, most of Mars's atmosphere is made up of CO_2. But with so little atmosphere to go around, Mars doesn't get a whole lot of greenhouse warming. Typical nighttime lows go down to −128 degrees Fahrenheit, while daytime highs only get as high as −24 degrees Fahrenheit.[53] Mars is definitely a place to chill.

It's also a desert. There's very little water in Mars's atmosphere—

just 0.01 percent of what's found in Earth's.[54] Since the atmospheric pressure is so low, exposed liquid water boils away in seconds. This is the same effect you get when you try to boil water high in the mountains—the water doesn't need to get very hot before it turns to vapor. That's why the water that *does* exist on Mars is either gaseous (water vapor) or locked up in ice at the poles. There may, however, be a lot of water underground, as ice or even in liquid form.

So, depending on which part of your spacesuit failed, conditions on Mars today would quickly kill you, either from asphyxiation or hypothermia. And yet, for all Mars's differences from Earth, the Martian climate machine still operates in ways very familiar to earthlings.

Imagine for a moment you are a Portuguese sailor in the 1400s. You're trying to get from West Africa, where you've been trading, back to Portugal. If you try sailing directly northward, you'll find storms and variable winds that move you along at a sluggish pace. But if you try something crazy and sail west—out deeper into the Atlantic and away from Portugal—you get a pleasant surprise. Sail far enough west and you hit beautiful, steady winds that will carry you back east *and* north. You're home in Portugal in no time. What you've discovered are the trade winds.[55]

A couple of hundred years after European sailors stumbled on the trade winds, English lawyer and naturalist George Hadley found their explanation. The trade winds are giant rivers of air, driven by solar heating and the Earth's rotation. Hadley recognized that hot air in the tropics always rises upward, while cold air at the poles always sinks. The air in between has to fill in the gaps, leading to a giant equator-to-pole pattern of circulation.[56]

If the planet weren't spinning, that would be the end of the story:

up/down and north/south motions. It's Earth's rotation that bends the equator-to-pole atmospheric conveyor belt through what's called the Coriolis force, which twists the flow, adding an east/west component to the circulation. The big circular flow in the North Atlantic is one of these giant rivers of air. In the southern hemisphere, there's a mirror-image trade wind pattern (the east/west direction is flipped because the direction of the Coriolis force changes across the equator). In total, Earth has six of these vast, circulating atmospheric flows, and the strongest of these, flowing just above and below the equator, are called Hadley cells.

Mars, like Earth, is spinning. At 24.7 hours, the length of its day is remarkably close to Earth's.[57] Since the laws of physics don't care where you live, Mars's earthlike spin should mean Hadley cells appear on the Red Planet, just as they do on our world. "It's one of the first things that comes out of a good Mars climate model," says Haberle. "You see big circulation patterns from the Martian equator to pole and back again."[58]

The Hadley cell is not the only familiar climate pattern on Mars. "Mars has jet streams," says Haberle, referring to the rivers of fast-moving air that exist high in Earth's atmosphere. "Every rotating planet with an atmosphere has them." And, just as on Earth, sometimes those jet streams will buckle and wander. Atmospheric scientists call these flow patterns "Rossby waves," and they were the cause of the dreaded "polar vortex" that brought record-cold air to inhabitants on the East Coast in the winter of 2014.[59]

While their technical details can be daunting, Hadley cells, jet streams, and Rossby waves all show us something profoundly simple and important: the physics of climate is universal. All worlds obey the same rules: Earth, Mars, Venus, even an exoplanet a hundred light-years away. Most importantly, they are rules that we now understand because we've seen them working on more than one planet.

HABITABLE WORLDS, SUSTAINABLE WORLDS

If you want to know the weather on Mars right now, there is an app for that.[60] The *Curiosity* rover, which landed in 2012, includes a meteorological station that beams conditions back to Earth for any and all to see. Follow the app for a whole day, and you'll see the temperature rise and fall between very un-earthlike extremes. You'll also see the atmospheric pressure change in ways that are definitely not witnessed on our world.

On any given day, the amount of atmosphere pressing down on the Martian surface can change by as much as 10 percent. That's almost like being in Los Angeles in the morning and then climbing the mile up to Denver's thinner air a few hours later, only to return again to sea level by nightfall. For our story, what's important about these changes is that the dramatic pressure swings are completely captured in the Martian climate models. There is so little air on Mars that, once the Sun begins warming the surface and driving hotter air upward, the entire planet's atmosphere readjusts, sending pressure waves from one side of the globe to the other. All the models track these readjustments and nail the daily air pressure variations. In other words, the climate models get these answers right.[61] That alone is an important point for our earthbound climate debate. We understand climate well enough to predict it on other planets.

But the success in understanding Mars's climate today highlights the single most important lesson Mars has taught us: climate changes, and with it, habitability changes, too.

Habitability is a key concept for astrobiologists, who think of it in an intuitive way: the ability of a planet to be inhabited by life. In the Drake equation, the formal definition of habitability is the presence of liquid water on a planet's surface.

The robots we sent to Mars offer us pretty definitive evidence that Mars once had liquid water on its surface. Some of that evidence is geological and comes via mineralogy. The exposed Martian bedrock in front of *Opportunity* not only sent Steven Squyres into fits of joy, eventually it also revealed the presence of small, spherical pebbles termed "blueberries." Instruments embedded in the tip of the rover's multi-jointed arm allowed Squyres and his team to recognize these blueberries as hematite, a mineral that only forms in the presence of liquid water.[62]

Some of the evidence for a wet version of Mars was more direct. Seven years after the discovery of the blueberries, the *Curiosity* rover found a set of carved rock features on Mars that could only have resulted from deep, fast-moving water flows. Curiosity scientists could even estimate the nature of the flow—about hip deep and rushing downstream at three feet a second.[63]

So, Mars once had liquid water on its surface. But that means it must also have had a much thicker atmosphere keeping that water from flashing away into vapor. And if the water was rushing on the surface, that thick atmosphere must also have been warming the planet to temperatures well above freezing. Put it all together, and it seems the Red Planet was once blue, at least for a while.

Scientists call this warmer, wetter period of Martian climate history the Noachian, for the story of Noah and the flood. Their best estimates place it between 4 billion and 3.5 billion years ago.[64] There remain deep questions about what happened to the water on Mars. Getting answers to those questions may have to wait until we can send actual geologists to the Red Planet.

But even as we wait for those answers, the recognition of Mars's dramatic climatic change already offers us a critical astrobiological perspective on our own Anthropocene era. Mars shows us that habitability—that most critical of astrobiological concepts—is not forever. A planet can change its habitable state. Most importantly, it can lose it entirely.

When we worry over our entry into the Anthropocene, we are inherently concerned with our project of civilization's sustainability. But what is sustainability but a special example of habitability? What we are really concerned with when we talk about the Anthropocene is the habitability of the planet for a particular kind of energy-intensive, globally interdependent, technological civilization. The present climate epoch—the Holocene—has been particularly habitable for that kind of project.

Mars shows us that habitability can be a moving target. The same is likely to be true of sustainability in the Anthropocene. Planets change, and that is a lesson Mars and its history help us come to terms with. It is not, however, the only lesson the other worlds in our solar system have to teach us.

REMARKABLE JOURNEYS

On June 12, 1982, Central Park hosted a sea of humanity. Spilling over the Great Lawn and onto Fifth Avenue, the park was thronged as never before in its 150-year history. The *New York Times* reported that there were "pacifists and anarchists, children and Buddhist monks, Roman Catholic bishops and Communist Party leaders, university students and union members." Delegations had arrived from Vermont and Montana, Bangladesh and Zambia. "The smiling, hand-clapping line of marchers threading into the park stretched back three miles along Fifth Avenue." According to the *Times*, it was "the largest political demonstration in US history."[65] All those delegations and all those people were in the park for one reason: to save the world.

The shadow of nuclear warfare, which loomed so large as the final factor in Frank Drake's equation, had grown longer and darker by the early 1980s. The election of Ronald Reagan as president, along with renewed aggressive actions by the Soviet Union, seemed, once again, to edge the world closer to an all-out nuclear exchange. By 1982, the two superpowers had increased their stockpile to more than fifty thousand nuclear weapons.[66] The massive rally in New York was intended to build support for a "nuclear freeze"—an end to the weapons buildup and the beginning of a nuclear drawdown. But neither the US nor Russian government was listening.

In response, a new kind of peace movement grew. It was larger and broader than anything the cold warriors of the 1960s had been forced to contend with. While the Central Park rally marked the nuclear freeze movement's rise to political relevance, its framing of humanity's basic nuclear dilemma differed significantly from that of the Cold War era twenty years before, when Frank Drake formulated his final factor. This shift became apparent a year after the massive rally, when a group of scientists published a study that changed the language of nuclear war.

The paper was titled "Nuclear Winter: Global Consequences of Multiple Nuclear Explosions." Carl Sagan and James Pollack were both on the team of authors who were collectively referred to as TTAPS (Richard P. Turco, Owen Toon, Thomas P. Ackerman, Pollack, Sagan). The TTAPS argument was straightforward: even a medium-scale nuclear exchange would lead to so many fires that soot lofted into the atmosphere would significantly cool the planet. Agricultural production would seize up and the world would be plunged into hunger and chaos. The lesson from their study was straightforward too: almost any nuclear exchange could transform the planet in dangerous ways. The weapons could never be used.[67]

By this time, Carl Sagan had become a celebrity via his best-selling books and his TV appearances. He highlighted the TTAPS study with an extended essay in *Parade* magazine.[68]

While the Reagan administration publicly dismissed the science of nuclear winter, the majority of the scientific community took it seriously. From that point, there was no going back. "Nuclear winter" entered the world's vocabulary and its imagination. Years later, both Soviet and American officials would openly discuss how nuclear winter's doomsday scenario helped draw the two nations to the negotiating table.[69]

The entry of nuclear winter into the political landscape was notable for two reasons. First, it was a result based on a climate model. Pollack, Sagan, and their collaborators had used the mathematical physics governing global atmospheric flows to track the behavior of particles blown into the air by global fires. For the first time in human history, a model of a planet's climate would frame global political debate. But it's a second feature of the debate that matters most for our moment. A key argument for nuclear winter came from Mars.

The globe-engulfing Martian dust storms, first observed in detail by *Mariner 9*, provided critical data for the nuclear winter researchers. The behavior of tiny particles carried high into the atmosphere would have been mere theory without the flotilla of probes we'd sent to Mars. With the data they supplied, the Martian climate models were expanded to include newly realized physical principles of how solar and infrared radiation interacted with dust. Thus, the space probes and the climate models revealed the powerful effect of dust on the Martian atmosphere. That understanding was then transferred to the distinctly terrestrial problem of fires ranging across the planet after a nuclear war. The TTAPS paper was explicit in calling Mars out as a test bed for nuclear winter physics.

The history of TTAPS and nuclear winter shows us that knowledge gained from an alien world has already influenced earthbound debates about our future. Now, thirty years later and in the midst of our modern climate debates, we must recognize how deeply our understanding of climate is rooted in what we've learned from "wheels-on-the-ground" studies of other planets. The desperate attempt by climate-change deniers to sow doubt on climate science (and its modeling efforts) willfully ignores what five decades of space travel have taught us: we have more than one world, and one story, to school us in the ways a planet can change.

We humans sent exemplars of our ingenuity to Venus and Mars. Later, humanity's robot emissaries would reach the outer worlds of Jupiter and Saturn (and their remarkable ocean-bearing moons). By 2016, every planet and every class of solar system object had been visited at least once by our probes. Asteroids, comets, and dwarf planets—we had "touched" them all and we had learned from them all.

In making those remarkable journeys, we did more than simply satisfy our curiosity or beat other nations for bragging rights. While we might not have known it at the time, these missions to other planets were also giving us the conceptual tools we now need to make fateful decisions about our own still-unknown fate.

We could not have understood the greenhouse effect as we do now without what was learned from the robot probes to Venus. We could not have understood the process of climate modeling as we do now without the rovers trundling across Mars. And the atmospheres of Jupiter, Saturn, and other solar-system worlds have each provided their own lessons. We traveled billions of miles only to see our planet and our own predicament come into high resolution.

CHAPTER 3

THE MASKS OF EARTH

AIR LESSONS

Imagine you are a time traveler who just landed on Earth 2.7 billion years ago. As you step outside on this younger version of our planet, what's your first experience? The answer is pretty simple. You die.

To be specific, you asphyxiate. For about the first two billion years of Earth's history, its atmosphere contained only minute traces of oxygen, even though it had long been a home to life. For almost half the planet's history, its "air" was composed almost entirely of nitrogen and CO_2.[1] Today, however, the Earth's atmosphere is almost all nitrogen and oxygen, with only a tiny fraction of CO_2. What happened to make so great a change?

This one all-important detail about Earth's history—the rise of its oxygen—is a lesson for us today. It was life, acting on a global scale billions of years ago, that altered the planet's atmosphere. In doing so, it also changed the future history of the Earth, leading to humans and our project of civilization. Now life, in the form of our civilization, is once again poised to alter the planet's atmosphere and its complex machinery of evolution. The comparison of that time, billions of years past, with our own moment of climate change offers a doorway into the remarkable story of the "masks of Earth." Its narrative bears a truth few of us recognize.

Our world has been many planets in the past.

These other versions of Earth were profoundly different from the cloud-mottled, blue-green world we know today. Each was a consequence of planetary forces shaping and then reshaping our world. Together, they reveal how deeply humans and our project are part of a much longer story. When it comes to life changing the planet, we are neither unique nor unusual. That's why the story of our planet's past, a story that is fundamentally astrobiological, is so critical to us. Knowing the Earths that *were* will give us the vocabulary to craft a new story, one that keeps us part of the Earth soon to be.

NO EASY DAY

The Polecat arctic transports were beasts. Designed for duty in the harshest conditions, each was the size of a minibus. The caterpillar-treaded special-purpose vehicles were built wide to keep them steady on uneven terrain, with powerful diesel engines for hauling cargo and personnel across ice, snow, or even up the steep side of a glacier.[2]

On October 16, 1960, the side of a glacier was all young Soren Gregersen saw as he looked out the window of his assigned Polecat. Just a few hours earlier, Gregersen, a seventeen-year-old Danish Boy

Scout, had been stuffed into the Polecat's cab by a smiling GI. Two days before that, he'd landed at the US Air Force's Thule Air Base on the western shore of Greenland and been outfitted with regulation cold-weather military gear. Gregersen watched in wonder as the Polecat began its long trek up the "ramp," a sloping road carved into the glacial ice. He was beginning a 150-mile trek out onto one of Earth's most inhospitable locales.[3]

Bouncing around the Polecat's cab, Gregersen was caught somewhere between excitement and terror. After all his hopes, preparations, and travel, it was really happening. He was on his way to Camp Century, a city the Americans built under the ice.

At the same time that Jack James was blasting Mariner probes to the planets and Frank Drake was tuning his radio telescope in search of alien civilizations, the US military was pushing audaciously across a different kind of boundary. This one, where Boy Scout Soren Gregersen was bound, lay at the top of the world.

Greenland is a giant ice slab where seven hundred thousand square miles of glacier rise a mile and a half above sea level.[4] Temperatures at the center of its vast ice plateau typically drop to a Marslike −70 degrees Fahrenheit. Winds routinely sweep across its barren plains of snow at 125 miles per hour.[5] And yet, in 1959, the US government chose to build a military base and a scientific laboratory right in the middle of Greenland's frozen emptiness.

The logic of the Cold War led the US to plan the impossible in the form of Camp Century. The base consisted of twenty-one trenches dug into the ice, each up to three football fields long. Each trench was twenty-six feet wide and twenty-six feet deep, with snow packed across steel arches to create a ceiling. Prefab buildings, hauled across the glaciers, had been laid into each trench to serve

The barracks built inside one of the ice
tunnels at Camp Century.

as barracks for the camp's two hundred servicemen and scientists.
Powering the base required a $5 million portable nuclear reactor
that the military dragged out across the ice sheet.[6] Taken together,
building Camp Century required a monumental effort, but one that
would achieve a monumental breakthrough.

In our era, when people who know nothing of climate science
make sweeping claims of sweeping ignorance, it's important to
remember the risks required to make that science happen. The soldiers and scientists at Camp Century lived at the edge of the world,
and their work carried considerable dangers. Transport flights and
crews had to contend with extremes of weather unlike almost any-

where else on the planet. In the summer of 1961, a helicopter crash outside the base took the lives of all six aboard.[7] But those GIs, their officers, the scientists, and even Boy Scout Soren Gregersen had all come to Greenland's glacial wasteland on a mission. "It was the most exciting thing that ever happened to me," recalled Gregersen, now a retired professor of geophysics, when I spoke with him. "That experience is what got me started in science."

Camp Century was a joint US–Danish effort (Greenland is a Danish territory). To create publicity for the polar mission, the Boy Scouts in both countries held competitions seeking "junior scientific aides." In late 1959, Gregersen and American Boy Scout Kent Goering each won their chance to spend five months on, and in, the ice.

"We lived right alongside the GIs," says Gregersen. "Every day, we got some task required to maintain the base. Sometimes it was chopping away the ice that constantly grew inside the tunnels. Sometimes it was working on the pumps that fed an enormous freshwater reservoir deep in the ice. I loved it all, and all of it was thrilling."

But for young Gregersen, it was the science that made the strongest impression. There were many reasons why the US military built Camp Century. Plans had been discussed to house nuclear missiles in the ice (the ever-shifting glaciers killed that idea).[8] There was also the need to keep watch on the Soviets. Gregersen remembers the vast radar arrays, pointing north, at Thule Air Base. But the military was especially interested in climate. The history of warfare was, after all, full of military campaigns done in by weather. Just as the funding for the exploration of space was opened by the Cold War, the Earth's climate and its history had also become a military concern. That translated into funding for climate science. The money took scientists to the most remote corners of the planet. It was also how young Soren Gregersen first saw the drills at Camp Century.

In rooms carved from centuries of fallen snow, Camp Century scientists set up drilling derricks, like the kind you'd see in oil country. Their goal was to dive downward through almost a mile of ice and thousands of years of planetary history.[9] "I saw the effort being made in those ice drilling labs," says Gregersen. "And it made a huge impression on me. What they were trying to do—it just seemed amazing—recovering the history of the planet using ancient snow."

Transformative visions of the world usually come when we find new ways to see it. In science, the ability to get to new kinds of data—literally new ways of seeing—allows us to revise and refresh our understanding. Jack James's Mariner mission to Venus, Carl Sagan's Martian dust data, and the radio telescopes at Frank Drake's Green Bank observatory all rewrote our understanding of astronomy and planetary science. In the decades after World War II, our understanding of the Earth was also being reimagined by new data that had been beyond the reach of earlier generations of researchers. Camp Century was one critical chapter in the story of that change.

Ice ages were still a mystery in 1960. The most certain thing scientists could say about them was that they'd happened. Over the last few million years, mile-thick slabs of ice covered much of the Northern Hemisphere. At least four different times, they ground their way south and then retreated back.[10] Each glacial epoch left the planet cold and dry. Ocean levels dropped almost four hundred feet—the height of a forty-story building—as so much of the Earth's water became locked in ice. In between the ice ages, the planet got reprieves in the form of warmer, wetter interglacial states.[11]

The Earth endured the last ice age for almost a hundred thousand years. Only after the final laggard glaciers retreated did the project of human civilization begin. Our history of farming and cities, writing and machine building fits entirely within the Holocene: the current ten-thousand-year-old interglacial period.[12] And even though scientists knew the basic sequence of events leading to the Holocene, the details of how the climate slipped from one state to another eluded them. They simply didn't have the data to see the details of the change. What they needed was a way to follow the planet's temperature, year by year, all the way back to when glaciers were last king. Under the auspices of the US Army's Cold Regions Research and Engineering Laboratory, Camp Century's drilling operation gave scientists that record.

The work was led by the Danish scientist Willi Dansgaard and the American geophysicist Chester Langway. The mile-thick slab of ice covering Greenland is maintained by yearly layers of snowfall, packed one on top of the other. The strata of ice, built up year by year over the millennia, form a kind of frozen layer cake. Each layer comprises a record of that year's climate. Within each layer of ice was a chemical marker that served as a proxy thermometer. Using it, scientists built a high-resolution recording of Greenland's temperatures going back thousands of years.[13]

After six relentless years of work, Dansgaard, Langway, and their Camp Century team drilled all the way down to bedrock, more than four thousand feet below the top of the ice sheet. Once the "ice core" data retrieved by the drilling was converted into temperatures, Dansgaard and his colleagues could see Earth's passage out of the last ice age. Moving backward, they first saw a period of roughly constant temperature stretching back eight thousand years. This was the Holocene, the time during which human civilization had been born and grown to thrive. Going farther backward, they could also see the transition from the warmth of our current climate to the frozen glacial age more than ten thousand years ago (the Pleistocene).[14]

Along with the smooth transition from the last ice age to the current warm interglacial period, the Camp Century data also showed a series of spectacular short-term shifts that would come to haunt our climate future. Around twelve thousand years ago, in a period called the Younger Dryas, the planet appeared to drop from a warming state back into the icebox. It was a stunning discovery. In just a matter of decades, average temperatures around the planet had dropped by five degrees Fahrenheit in some places and as much as twenty-seven degrees in others.[15] If comparably dramatic global changes occurred in the modern era, it's hard to imagine our project of civilization making it through intact.

Later drilling work in Greenland and Antarctica confirmed the Camp Century studies. One American researcher working in Antarctica recalls a moment of truth when just looking at an ice core

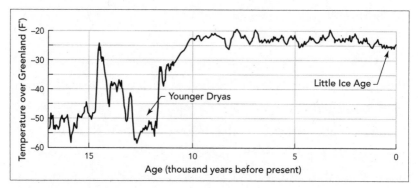

The history of Greenland temperatures based on ice core records.

made the speed of climate change apparent. The ice changing from light to dark across just a few inches in the core was a visceral confirmation of abrupt large-scale swings in global climate.

The recognition of rapid climate change presented researchers with a warning the importance of which they could not yet understand. At the time, human-driven, or "anthropogenic," climate change was nothing more than a possibility discussed in the most abstract terms at meetings of scientific experts. Almost no one was ready to conclude that the kind of rapid climate shift seen twelve thousand years ago might be something we could drive through our own actions.

WHICH EARTH?

Whatever quiet preparations were going on in homes across the Earth, William Anders was not part of them. That's because Anders was on a spaceship. On Christmas Eve 1968, two hundred thousand miles from the planet of his birth, Anders and fellow *Apollo 8* astronauts Frank Borman and James Lowell were becoming the first humans to orbit the Moon.

"Oh my God," Anders said to his crewmates as he marveled at the view outside the small window of his Apollo command module. "Here's the Earth coming up," he said, looking out across the moon's horizon. "Wow, is that pretty."

Anders asked for a roll of color film while Borman joked, "Hey don't take that [picture], it's not scheduled." Loading up the camera, Anders stopped for a moment to consider the magnitude of

The iconic *Earthrise* photograph taken by William Anders during the *Apollo 8* mission in 1968.

the vision before him. Then he snapped an image of the world that would change the world.[16]

Called *Earthrise*, Anders's picture of the blue Earth hanging above the gray moonscape became iconic. *Life* magazine named it one of the one hundred most influential images in human history.[17] Since then, space-based pictures of azure oceans, swirling white clouds, and green-brown continents have become familiar. But that familiarity is undercut by a striking truth that has been emerging since the time of Camp Century: the planet we know today is not the Earth that was. If you had visited our world one hundred million, five hundred million, or three billion years ago, you would have found a planet that looked very different from Anders's image.

Exhaustive work going back to the 1800s has allowed geologists and paleontologists to construct a timeline of our world's history. But only in the last half century or so has that timeline been resolved into the details of planetary change. There are four long *eons* in the Earth's history, representing the most important transitions in the planet's climate and life. These eons are subdivided into eras, which are further divided into periods and epochs. The Pleistocene and Holocene, whose transitions were revealed by Camp Century ice cores, are examples of epochs.[18]

The planet's story begins with an unnamed cloud of interstellar gas and dust. Almost five billion years ago, that slowly spinning cloud, close to a light-year across, collapsed under its own weight. The Sun formed at the center of the infalling mass, and a rapidly spinning disk surrounding the young star emerged as well. Within this dense disk, particles of dust began colliding frequently enough to form free-floating pebbles. Those pebbles then collided to form rock-sized objects. The rocks then collided to form boulders, and so on, all the way up to asteroid-sized *planetesimals*. After between ten million and a hundred million years, gravity drew the planetesimals together and assembled the Earth and other rocky planets (Mercury, Venus, and Mars).[19]

This was the beginning of the Hadean, Earth's first eon. Lasting from 4.6 billion to 4 billion years ago, its name speaks to the planet's hellish conditions. Earth during the early Hadean was covered in a globe-spanning sea of molten rock. Eventually, this magma ocean cooled and hardened, forming a solid surface. But asteroids and comets continued to rain down on the planet, ending in a period called the Late Heavy Bombardment, when our solar system cleared itself of planetary construction debris. Each of these apocalyptic impacts shattered the surface, turning some or all of it back into molten rock. Gases released from the bombardment and the magma oceans it regenerated left the Hadean Earth with an atmosphere composed mostly of nitrogen and carbon dioxide.[20]

Thus, the Earth was once a fire world of molten seas.

The planet's first forms of life may have emerged by the end of the Hadean. The repeated asteroid bombardments would, however, have sterilized the world, forcing biology to potentially start over and over again.[21] Either way, by the beginning of the next eon—the Archean—the kind of life we know today was already in place. The Archean lasted from 4 billion to 2.5 billion years ago. It was during this vast span of time that life based on the biochemistry of self-replicating molecules called DNA spread across the world. But in the Archean, all life consisted of simple, single-celled organisms living in the oceans. The reason for this watery fixation was simple: the whole planet was pretty much an ocean.[22]

While continents now cover about 30 percent of the Earth's surface, during the Archean they had yet to "grow." The ground you stand on today is composed of granite that is less dense than the black volcanic basalt making up the ocean floor. Granite is formed deep within the Earth's mid-layers (called the *mantle*). Like warm air in a cold room, granite rises slowly upward as it forms, allowing it to become separate from the more dense ocean crust. While there remains controversy about the process, many scientists believe that during the Archean the continent making was still beginning. Rather than planet-spanning continents, the world hosted just one or two proto-continents called *cratons*. Each craton was smaller than India is today.

Thus, the Earth was once a water world of almost endless ocean.

Life slowly explored new domains of structure and metabolism as the Archean gave way to the Proterozoic eon, lasting from 2.5 billion to half a billion years ago. The earliest cells on Earth had been relatively simple affairs. Called *prokaryotes*, they include modern-day bacteria. The first prokaryotes lived by breaking down complex molecules into simpler structures (basically fermentation). The evolution of early forms of photosynthesis had, however, given some prokaryotes the ability to draw energy directly from sunlight. These were the earliest forms of photosynthesis, whereby cells use sunlight to generate food.[23]

By the beginning of the Proterozoic eon, life had learned new, more efficient photosynthetic strategies. Some of these came from the development of a wider range of internal machinery, like a cellular nucleus to hold the genetic blueprints of the cell. The emergence of these nucleus-bearing eukaryotic cells changed life's trajectory on the planet. With the addition of new forms of photosynthesis, more energy became available to cells, allowing them greater flexibility and adaptation. The first multicellular organisms appeared during the Proterozoic, as life began to experiment with the division of labor. Cells specialized into different forms that worked together. Left without the larger organism, however, these specialized cells would die.[24]

Along with the changes in life, the planet itself was changing. During the billion-year-plus stretch of the Proterozoic, the first cratons grew into full-sized continents. Eventually, the slow movement of the Earth's crustal plates (plate tectonics) drew them together to form a supercontinent, a vast landmass called Rodinia. Other supercontinents would form and break apart over the course of Earth's history. Each would change the planet's climate by altering global ocean circulation and resetting patterns of rock weathering and CO_2 cycling.[25]

Perhaps the most important climate shifts to come during the Proterozoic were the first periods of near-total glaciation. At least four times during this eon, changes in the concentrations of atmospheric greenhouse gases plunged planetary temperatures into the freezer. From the poles all the way to the equator, the entire planet may have become locked in miles-thick layers of ice.[26] Seen from space, this snowball world would have appeared as a mottled and cracked Ping-Pong ball with no large expanses of open blue water.

Thus, the Earth was once a snowball world of endless ice.

For all the changes Earth experienced, none was more remarkable or mysterious than life's sudden burst of creativity just after the Phanerozoic eon began 540 million years ago. Across a remarkably short span of geological time, evolution threw itself a party. What began as still-simple multicellular life rapidly diversified into an orgy of new forms and new species. In just fifty million years, evolution produced all the basic structures that mark life on Earth today. Called the Cambrian Explosion (it occurred during the Cambrian geologic era), it was an evolutionary acceleration on a scale never seen before or after.[27]

It was only after the Cambrian era that all the "prehistoric" worlds we know from popular fictions arose. There was the Carboniferous era three hundred million years ago, with its vast swamp forests. Those forests eventually became the coal beds we've used to power our project of civilization.[28] There was also the Jurassic era, dominated by the huge dinosaurs that live on in movies and the dreams of little kids. And finally, there was the more recent cycling of ice ages and interglacial periods, during which we humans appeared and eventually flourished.

The Earth swung back and forth between many versions of itself during the fecund eon of the Phanerozoic. But of particular interest to our own age are the periods when the planetary thermometer rose to fever levels.

Fifty-five million years ago, the supercontinent called Pangaea began splitting apart. The volcanism that accompanies plate tectonics went into overdrive, dumping CO_2 into the atmosphere far faster than it could be removed by natural feedbacks. Global average temperatures rose fourteen degrees Fahrenheit above what we experience today. Called the Paleocene-Eocene Thermal Maximum, the result was a planet almost without ice.[29] Temperatures in Greenland, where a future Soren Gregersen would endure his subzero glacial summers, stayed at a balmy 70 degrees Fahrenheit.

Thus, the Earth was once a jungle world, a sweltering hothouse planet devoid of snow.

Given the scale of the Earth's changes between one mask and another, the next question we should ask seems clear. What force was powerful enough to drive our world's dramatic transformations?

THE GREAT OXIDATION EVENT

The engineer asks Donald Canfield if he is claustrophobic. Canfield, a professor of ecology, has just squeezed himself into the cramped confines of *Alvin*, the world's most famous deep-sea submersible. It's a fall day in 1999 on a research ship slowly rolling in the Gulf of California's blue waters.

"Claustrophobic? No, not at all," Canfield says, lying enough to make them both feel better.

The engineer flashes him a knowing smile and says, "Good . . .

whatever you do, don't touch the red handle. It's only for emergencies."[30] The hatch slams closed.

After an hour-long descent, Canfield is skimming along on the floor of the Guaymas Basin in the Gulf of California, more than fifty miles east of the Baja Peninsula and over a mile below the surface. The basin is a "spreading zone" where two of Earth's continental plates are pulling apart.[31] As the plates move away from each other, they carry the Baja Peninsula away from mainland Mexico at a rate of about one inch per year, the same rate as your fingernails grow.[32] In between the spreading plates, new seafloor crust is constructed as hot magma upwells from deeper within the planet, cools, and then hardens into solid rock.

From the circular observing port cut into *Alvin's* six-foot titanium crew capsule, Canfield gets his first view of the basin's floor. Far from the well-lit upper ocean, it's an alien world laid out before him.

"All around us," Canfield recalls in his book, *Oxygen*, "We see the effervescence of hot [sulfur]-rich, hydrothermal waters percolating from the accumulating crust." Boiling water, heated by the Earth's internal fury, rises in dark columns from the vents. High-temperature geology is, however, only one facet of the otherworldly vision in front of Canfield. Remarkably, life is thriving here in the heat and the darkness. "Great mounds of *Riftia* tubeworms rise from the shadows swaying gently on expansive hills of gypsum crust," he writes.[33] The enormous tubeworms have no color—none is needed in this world of perpetual darkness.

Everywhere, Canfield makes out what appears to be fallen snow on the gypsum-crusted seafloor. What he sees, however, is not snow, but bacteria. The abundant microscopic creatures draw their energy from the heat and sulfur-based compounds spilling from the hydrothermal vents.[34] Their ability to thrive in such an extreme environment is what allows the whole strange ecosystem laid out before Canfield to exist.

Canfield made this trip to the ocean floor to gain insights into the Earth's past in terms of an alternative biochemistry. What he found at the bottom of the Guaymas Basin were hints pointing to versions of life that need no sunlight. These are, perhaps, vestiges of an early incarnation of the planet before its most significant trans-formational event: the rise of oxygen in the Great Oxidation Event.

"Try to imagine something so profound, so fundamental, that it changed the whole world," Canfield writes. "Think of something so revolutionary, that it forever changed the chemistry of the atmo-sphere, the chemistry of the oceans and the nature of life itself."[35]

After posing this question, Canfield surveys the critical moments in human history: the Great Plague, the Renaissance, and World War II. "These were important events," he writes. "But their influence out-side the human realm was small." He then goes on to consider the extinction event sixty-five million years ago that killed the dinosaurs, and the one 250 million years ago that took down almost 95 per-cent of all animal species on the planet. Even those events pale in comparison to Canfield's target. "Each of these major extinctions changed the course of animal evolution, but still, they did not fun-damentally alter the fabric of life or surface chemistry of Earth."[36] What, he asks, *did* so completely transform the Earth? The answer to Canfield's question turns out to be as simple as drawing a breath.

During Earth's earliest eras, much of biology may have been powered by chemistry akin to what Canfield saw on his dive. By the middle of the Archean, however, at least some single-celled organisms had figured out how to tap a new and abundant energy source: sunlight. The first emergence of photosynthetic organisms in the form of what scientists call *anoxygenic phototrophs* (non-oxygen-producing sunlight eaters) was a major innovation in the history of life. Through the remarkable trial and error of evolution (and lots of time), some bacteria developed molecular light receptors. These were nanoscale machines that absorbed energy from the Sun and used it to power chemical reactions that popped out sugar molecules. Sugar, in whatever form, is the basic chemical battery for all the metabolic shenanigans cells need to stay alive.[37]

After a billion or so years of non-oxygen-producing photosynthesis, nature got very creative. Sometime in the late Archean, evolution produced a new version of photosynthesis that, for the first time, used water to drive its chemistry. Because water is superabundant on Earth, cells using this new kind of photosynthesis won out over the older forms. But these organisms—called *cyanobacteria*, or blue-green algae—did more than just multiply. Sucking in water, CO_2, and sunlight, they also started spitting out molecules of oxygen as a kind of waste product of their activity.[38] In this way, their innovative water-eating, light-powered, oxygen-producing metabolism led them to become the most powerful force in the history of the planet.

Over time, the activity of the cyanobacteria dumped so much oxygen into the oceans and atmosphere that the entire planet was forced to respond. The geologic record shows evidence of early "whiffs" where atmospheric oxygen levels increased by small amounts. But by 2.5 billion years ago, the fix was in. Across just a few hundred million years, the concentration of atmospheric oxygen increased by a factor of a million.

This was the Great Oxidation Event, or GOE. Ironically, the rise in oxygen was poison to the bulk of the life that existed at the time. Oxygen's ability to bind with so many chemicals means it can quickly degrade the function of cells and kill them. But evolution figured out how to make lemonade out of lemons. It learned to work with oxygen's juiced-up chemistry to create better, more energetic forms of life. Soon, creatures that breathed in oxygen had evolved. They used the element to power faster and more complex metabolisms.[39] The big brain you're using to read and comprehend these words would never have been possible without oxygen's kick to evolution.

By the end of the GOE, the anoxygenic phototrophs, once the planet's masters, had been forced into oxygen-free warrens, learning how to live in places like the fetid sulfur pits of Yellowstone or even deep in our stomachs. In this way, the new oxygen-breathing forms of life inherited the open sea and open sky.

The presence of oxygen in the atmosphere also allowed life to colonize the land en masse. Before the GOE, cell-damaging ultraviolet radiation from the Sun (the kind that gives you sunburn) streamed unremittingly through the atmosphere. Only in the oceans, below the surface, was life safe enough from UV light to form rich ecosystems. But with oxygen came the atmospheric ozone layer. Ozone is a gas, made up of molecules with three oxygen atoms, that forms high in the stratosphere. It's a potent absorber of UV radiation. This ozone sunblock shield, which made the land safe for life, could not have formed without the rise in atmospheric oxygen.[40]

So, what does the GOE, with all its power and reach, teach us about the Anthropocene? It demonstrates that life is not an afterthought in the planet's evolution. It didn't just show up on Earth and go along for the ride. The GOE makes it clear that, at an earlier point in Earth's history, life fully and completely changed the course of planetary evolution. It shows us that what we are doing today in driving the Anthropocene is neither novel nor unprecedented. But it also tells us that changing the planet may not work out well for the specific forms of life that caused the change. The oxygen-producing (but non-oxygen-breathing) bacteria were forced off the Earth's surface by their own activity in the GOE.

So, from the GOE we gain insights that are themselves a turning of the wheel in humanity's conception of itself and its place in the cosmos. We come to an idea that touches both the deepest levels of scientific consequence and the highest forms of mythic understanding. We come to the moment where the biosphere, and our place in it, can be fully imagined.

THE BIOSPHERE BEGINS

Scientists become famous for a lot of reasons. There are those like Einstein and Darwin whose visions shatter old ideas. Their names go on to live forever in the pantheon of genius. Then there are those like Carl Sagan and Stephen Hawking, both brilliant researchers, whose talents as writers allowed millions of non-scientists to understand the beauty and power of science. But how many people have ever heard of Vladimir Ivanovich Vernadsky? His name is far from a household word outside of his native Russia. But that obscurity is destined to change along with the planet.

It was Vernadsky's ideas—and their genius—that heralded a new scientific conception of life's planetary context. As we enter more deeply into the Anthropocene, we will find Vernadsky already there, waiting for us to catch up with him.

Vernadsky was born in 1863 in the St. Petersburg of Imperial Russia. His mother came from nobility; his father was a professor of political economy and statistics.[41] Vernadsky's parents were known for their devotion to democratic and humanistic ideals. From them, he inherited a fierce determination to live by those ideals, which was grafted onto a love of science. Across eighty-two years of wars, revolution, and acute political turmoil, Vernadsky did not waver in his devotion to scientific inquiry. And even at the greatest personal risk, he never wavered in working for the freedom to pursue scientific ideas, wherever they led.[42]

Vernadsky began his scientific work in the chemical study of minerals. Traveling across Europe in the late 1800s, he was keen to apply the most modern methods of physics to the study of rocks. His goal was to bring precision tools to bear on questions about the planet's history. But even as Vernadsky was committed to exacting empirical studies, he was always more than a specialist. Across his career, he struggled to see how the whole emerges from the narrower stories scientists can unlock from the parts.

Russian scientist Vladimir Ivanovich Vernadsky.

In this way, Vernadsky built a solid, data-driven foundation for a new field called geochemistry, which unpacked Earth's history by examining the microscopic composition of its physical constituents. Then Vernadsky went further. It wasn't just geology and chemistry that were linked. In his eyes, biology also had to be brought into the planet's story at a fundamental level, so he initiated a second field: biogeochemistry.[43]

Vernadsky was often critical of biologists for the way they treated "organisms as autonomous entities." In his eyes, any individual species carried more than just an imprint of the environment within which it had evolved. Instead, the environment was shaped by the activity of life as a whole. As he put it, "An organism is involved with the environment to which it is [has] not only adapted but which is adapted to it as well."

This attention to both microscopic and macroscopic views led Vernadsky to his most important addition to the language of life in the context of its planetary host. Building on discussions with the Swiss geologist Eduard Suess, Vernadsky proposed that the study of the Earth would not be complete without understanding the central role of life as a planetary force. Earth, in his view, cannot be truly understood without understanding the dynamics of its biosphere.

Living as we do after astronaut William Anders's *Earthrise*, it's hard to imagine that the biosphere could ever be a new or radical idea. But it was Vernadsky who gave the concept its scientific birth. It was Vernadsky who first clearly articulated what later scientists, studying everything from the Great Oxidation Event to modern climate change, would slowly—and with great effort—come to verify: Life was not just a patchy green scruff holding a tenuous position between rock and air; instead, it was a planetary power as important as volcanoes and tides. It was an active force shaping the complex multibillion-year history of the world. As Vernadsky wrote in 1926:

> The matter of the biosphere collects and redistributes solar energy, and converts it ultimately into free energy capable of doing work on Earth. . . . The radiations that pour upon the Earth cause the biosphere to take on properties unknown to lifeless planetary surfaces, and thus transform the face of the Earth.[44]

Over the whole of his celebrated career, Vernadsky continued to modify and extended his concept of biosphere. Specifically, he saw it as a region—a shell—extending from below the Earth's crust (the *lithosphere*) all the way to the edge of the atmosphere. Within this shell, the action of life dramatically changed flows of matter and energy.

Most important for our own moment, Vernadsky saw that the world-shaping powers of life were both ancient *and* ongoing. "Adjusting gradually and slowly, life seized the biosphere," he wrote. "This process is not yet over."

It's the scale of his vision that makes Vernadsky so important to our story. Earth's entry into the Anthropocene is, at one level, purely an issue of interacting planetary processes. *Our* entry into the Anthropocene, however, is different. For us, it's also an issue of making meaning, of making sense of our place within the web of life that is also a force shaping the planet. Vernadsky envisioned a global view that achieved both. It was both scientific and mythic in scale, long before satellites and space missions could make such a global view of Earth tangible.

After his death in 1945, the limits of the Cold War meant it would take some time for Vernadsky's radical view of life and its planetary reach to reach beyond Russia.[45] But in time, Vernadsky's vision did find its champions. As human culture was reshaped by its new space age, two scientists in particular would pick up Vernadsky's biospheric vision and grow it into a full-fledged science.

BIOSPHERE RISING

James Lovelock was always the outsider's insider. From the first radio set he cobbled together as a boy in England after World War I, Lovelock was an inventor of prodigious talent. Eventually, that talent drew governments and corporations to seek his help.

During World War II, Lovelock's degree in chemistry took him into medical research, where he invented everything from precision airflow meters for studying the common cold to specialized wax pencils that could write on wet test tubes. This talent as a "maker" would eventually bring a degree of independence as his inventions drew a steady income. In the 1950s, Lovelock designed a cheap, portable device for detecting minute amounts of chemical contaminants. The patent was so valuable it allowed him to pursue science on his own terms, independent of an academic or government affiliation. But governments were still keen to sign him on to their projects.[46]

In 1961, Lovelock found himself at the same Jet Propulsion Laboratory in Pasadena, where Jack James and his team were exhausting themselves on the Mariner Venus mission. For Lovelock, the sprawling campus had the look of "a hasty airport with prefabricated cabins dotted over the hillside."[47] JPL had paid for his trip to its nascent campus because they needed his aid in designing sensitive instruments for the new space missions. Eventually, Lovelock was put on a team proposing experiments to search for life on Mars.

Sitting through meetings where biologists laid out plans to detect Martian microbes, Lovelock found himself unconvinced. "The flaw in their thinking," Lovelock recalls in his biography, "was their assumption that they already knew what Martian life was like. . . . I gathered the distinct impression that they saw it as like life in the Mojave Desert."[48]

But Lovelock, with an outsider's perspective that would haunt him later, came at the problem from a different direction. "I think we need a general experiment," he told the group, "something that looks for life itself, not the familiar attributes of life as we know it on Earth."[49] Pressed by the program manager to propose experiments that looked for "life itself," Lovelock was taken down a road that would lead him straight into the realms of Vernadsky's biosphere.

Lovelock's background in physics, chemistry, and biology led him to see the problem in terms of planetary atmospheres. He knew that life was keeping the air oxygen-rich. Take the biosphere away, and that oxygen would chemically combine with other compounds such that, if you waited long enough, the Earth's atmosphere would become oxygen-free. Without life, it would return to a state of "chemical equilibrium" dominated by the CO_2 released from volcanoes.[50]

Based on what he saw on Earth, Lovelock reasoned that life would always keep a planetary atmosphere in a state far from equilibrium. That meant the activity of life would constantly push on the planet's chemistry. The biosphere's continual resupply of oxygen, an element that would otherwise react away, was just one example of such a push.

Over the next two years, Lovelock continued to visit JPL and continued to work out the details of his atmosphere-as-life-detector experiment. But then, in September of 1965, a flash of insight showed him there was more to his idea than just an experiment.

In an office he shared with none other than a young Carl Sagan, Lovelock was poring over new data showing that the Martian atmosphere was dominated by CO_2. Unlike Earth's blanket of gases, Mars's atmosphere was locked into the same kind of dead chemical equilibrium as that of Venus. A CO_2-dominated atmosphere is exactly what you'd expect as the result of chemical reactions that were allowed to run their own course, like mixing a bunch of compounds together in a box and leaving the whole thing alone forever. It was at that moment that Lovelock saw the light.

"It came to me suddenly, just like a flash of enlightenment, that [for the chemistry of the Earth's atmosphere] to persist and keep stable, something must be regulating [it]." The identity of this "something" came to Lovelock just as quickly as the question had. "It dawned on me that somehow life was regulating the climate as well as the chemistry. Suddenly the image of the Earth as a living organism able to regulate its temperature and chemistry at a comfortable steady state emerged in my mind."[51]

It was a powerful image. Lovelock saw the Earth as a single entity—"alive" in some sense—and regulating itself in the same way our bodies maintain their temperatures. Lovelock soon began fleshing out the details of his idea, looking for specific mechanisms life could harness to adjust conditions across an entire planet. As the work progressed, he realized he needed a name for the idea. He thought to call it the "Self-regulating Earth System Theory," but a conversation with a neighbor, the novelist William Golding (author of

Lord of the Flies), convinced him otherwise. Golding suggested Lovelock name the theory after the Greek goddess of the Earth, Gaia.[52]

There is some irony in the fact that Carl Sagan, who did so much for our concept of Earth in its cosmic context, would be present for the insight that gave birth to Gaia theory. Given that Sagan was never very supportive of Lovelock's idea, it is even more ironic that he'd serve as midwife to the next crucial step in its development.

In the years following their divorce, the biologist Lynn Margulis almost single-handedly forced the scientific community to recognize the importance of cooperation, rather than just competition, in evolution. Her theory of endosymbiosis demonstrated how the tiny chemical-processing plants in our cells called *organelles* had once been independent organisms. Margulis proved that organelles—like mitochondria, for example—had been absorbed into larger bacteria billions of years ago to form a cooperative, symbiotic whole. This symbiotic evolution was likely the origin of the eukaryotic (nucleus-bearing) cells that transformed life's trajectory during the Archean eon.[53]

In the early 1970s, Margulis had become interested in the question of atmospheric oxygen and its microbial origin. When she asked her ex-husband, Carl Sagan, if he knew someone who might be good to talk with about the problem, he suggested Lovelock. From this unlikely introduction, Lovelock and Margulis began a collaboration that fully defined the Gaian concept of life as a self-regulating planetary system. Where Lovelock brought the top-down view of physics and chemistry, Margulis brought the essential bottom-up view of microbial life in all its plentitude and power.[54]

The essence of Gaia theory, as elaborated in papers by Lovelock and Margulis, lies in the concept of feedbacks that we first encountered in considering the greenhouse effect.

James Lovelock and biologist Lynn Margulis in front of a statue of Gaia.

The temperature of the human body always hovers around 98.6 degrees Fahrenheit. That's what is known as a steady state. At your death, your body drops back to room temperature. That's equilibrium. The same ideas can be applied to the oxygen in the atmosphere. The current levels of oxygen are held in a steady state by chemical reactions driven by the presence of life. But how does life keep the oxygen levels steady? We've already seen how photosynthetic bacteria gave Earth its oxygen-rich air. But why did oxygen levels rise up to 21 percent, and no further? This is an important question, because if the concentration of oxygen in the air were to climb as high as 30 percent, the planet would become a tinderbox. Any lightning bolt would trigger fires that wouldn't stop. So, what kept oxygen levels from rising above this dangerous threshold? To answer that question, Lovelock and Margulis turned to the idea of feedbacks.

In their Gaia theory, Lovelock and Margulis argued that life as a whole exerted global negative feedbacks on the planet. Those feedbacks had kept the planet in a series of long-term steady states over its history that were always optimal for making the planet habitable and inhabited. In other words, life kept the planet cozy for life. If, for example, oxygen levels rose too high, then the increased oxygen would, itself, trigger blooms of microorganisms whose biochemistry would lead to those levels being drawn back down. It was a very big idea indeed.

Lovelock and Margulis were offering a scientific narrative whose ties to the scale of world-building myth were explicit. It was Vernadsky on steroids, a vision of planetary evolution where life was not just a force, but a force with its own kind of intention. But just because an idea is big and beautiful doesn't mean it's true. In particular, with the all-important idea of intention—meaning life's intention with its Gaian feedbacks—the two scientists opened a Pandora's box.

THE BIOSPHERE BOUND

Oberon Zell-Ravenheart (whose given name is Timothy Zell) never met a New Age idea he didn't embrace. He is a pagan and a shaman. A fully ordained priest in the Fellowship of Isis, he also finds time to work as an initiate in the Egyptian Church of the Eternal Source. Zell-Ravenheart is also a Gaian, and that, in a nutshell, is why so many scientists hated Lovelock and Margulis's big, beautiful idea.

Early on, Gaia found itself scorned as a scientific theory by scientists but wildly popular in the larger culture. As historian and philosopher Michael Ruse puts it: "[The public] embraced Lovelock and his hypothesis with enthusiasm. People got into Gaia groups. Churches had Gaia services, sometimes with new music written especially for the occasion. There was a Gaia atlas, Gaia gardening, Gaia herbs, Gaia retreats, Gaia networking, and much more."[55]

Gaia theory came along just as the environmental movement and post-'60s New Ageism were going mainstream. In 1979, nuclear power became a national issue thanks to the partial core meltdown at the Three Mile Island generating station near Harrisburg, Pennsylvania. The pollution-driven evacuation of Love Canal in Upstate New York became the poster child for what environmental degradation looked like. Gaia theory, with its evocation of Earth as a single living organism—a vast planetary mother—channeled popular ecological concerns with an alternative vision of humanity's place in the scheme of things.

Many scientists pounced on Lovelock and Margulis for promoting the equivalent of snake oil. As microbiologist John Postgate, a fellow of the Royal Society, put it: "Gaia—the Great Earth Mother! The planetary organism! Am I the only biologist to suffer a nasty twitch, a feeling of unreality, when the media invite me yet again to take it seriously?"[56]

The real problem with Gaia theory for many scientists was the issue of teleology. It's a hallmark of biology that evolution that has no purpose, direction, or goal (*telos* is the Greek word for "goal"). The idea that the biosphere was somehow manipulating the chemical and physical conditions on the planet for its own good seemed inherently teleological (that is, goal-oriented). It smacked of intention, and evolution doesn't have intention.[57]

Lovelock and Margulis were unbowed in their defense of Gaia.

In response to critics who claimed their proposed feedbacks were nothing more than fantasy, Lovelock produced his now-famous Daisyworld model. Developed with mathematician James Watson, the Daisyworld model used a simple set of equations to describe a planet with two species of daisies (black and white) and a gradually brightening sun. The solutions to the equations showed clearly how feedbacks from the daisies (the black ones absorbed sunlight, while the white ones reflected it) could naturally keep the planet at a steady temperature even as the sun heated up. It was a tour de force of representing a complex idea with simple math in the service of proving an essential point. As Lovelock put it, "Daisyworld keeps its temperature close to the optimum for daisy growth. There is no teleology or foresight in it."[58]

And Lovelock and Margulis made it clear, they were not claiming the planet should be considered alive in any true sense of the word. New Age Gaian Mother Earth ceremonies notwithstanding, Lovelock and Margulis were ultimately arguing for the central role of the biosphere in planetary evolution. They were picking up where Vernadsky left off, and putting in more science.

With the publication of the Daisyworld model in 1983, the tide, at least partially, began to turn. Biospheric feedbacks were recognized as an essential part of planetary laws of operation. These feedbacks represent the definition of how to think like a planet, and researchers embraced the biosphere's central role in their studies of the Earth. But in the process, the name "Gaia theory" was dropped and replaced with the less contentious "Earth system science." While the concept of self-regulation remained contentious, researchers now knew that the linkages between the biosphere, atmosphere, and other systems were so tight that they had to be considered together as a single entity. The adoption of the Earth system paradigm represented its own revolution in how we think of planets, and today it forms the cross-disciplinary foundation for all researchers trying to understand climate change.[59]

As these studies of Earth system science were extended to include the planet's past, a crucial new idea would be added to the researcher's lexicon. Building on Vernadsky, Lovelock, and Margulis, a new generation of scientists began speaking of a "coevolution" between life and the planet. That word, coevolution, would help rewire astrobiology. Life could no longer be isolated from the planet that gave it birth. Instead, a planet could be deeply transformed by the life it births, including when that life goes on to create its own globe-spanning civilization. Thus, within that single term, coevolution, lay the seeds of a new story waiting to be told about humanity and our Anthropocene.

CHAPTER 4

WORLDS BEYOND MEASURE

HOW TO RUIN YOUR LIFE WITH PLANETS

A ll of Thomas See's fellow astronomers hated him. That was a particularly ironic position for See to find himself in, given that he was one of the most popular of "popular" astronomers in the late nineteenth century.[1]

See began his career full of promise. He was considered an expert with a telescope, and his skill as a writer for non-scientific audiences made him the astronomer reporters turned to when they needed a quote or an explanation. But his meteoric ascent would later be followed by a fall into the depths of scientific scorn. In the end, See was so despised by his colleagues that his experience became an object lesson in ruining a scientific reputation.

It's a story that begins with a planet.

See was born in rural Missouri in 1866. Though he was clearly a gifted child, his family would not allow him to attend school full-time until his teens. Once there, his natural aptitude for science and mathematics caught the attention of teachers who helped him get into the state university. Later, his natural talent led him to work with some of the best astronomers of the era, studying pairs of orbiting suns called binary stars.

See's work involved precision mapping of the sibling stars as they changed position on the sky. He was tireless in these "astrometric" studies. He worked eighteen-hour days, translating the information in photographs produced over many nights of telescopic observation into positions on sky maps. This astrometric data was then fed into calculations that spit out the exact shape of the binary stars' orbits. Finally, estimates of the stars' masses could be extracted from the orbits by using laws of physics. No one knew very much about how much mass stars contained back in the 1890s, and See's work was hailed as cutting-edge science.

See was hired at the University of Chicago, and then at the observatory that Percival Lowell, the rich, Mars-crazy amateur

Astronomer Thomas Jefferson
Jackson See.

astronomer, was building in Flagstaff, Arizona. It was at Lowell's observatory that the trouble began.

In 1899, See published a letter in the prestigious *Astronomical Journal* claiming the binary system called 70 Ophiuchi was "perturbed by a dark body." He meant the orbits of the two stars seemed to be distorted by the gravity of a third, unseen object. Later, See would claim to see other binary stars with invisible companions, reporting, "They seem to be dark . . . and apparently shining by reflecting light. It is unlikely that [the unseen objects] will prove to be self-luminous." See was being coy with his language, but the implications of his statement were unambiguous. He was telling the world he'd discovered other planets orbiting other stars.

The question of whether other stars in the sky might have planets goes back to the ancient Greeks. For millennia, astronomers and philosophers argued about the existence of other solar systems in the cosmos. Giordano Bruno risked his life arguing that other worlds exist. That is why direct evidence for the existence of even one planet orbiting one other star would have been an epoch-making discovery. See was making an extraordinary claim with his orbit-perturbing "dark body" of a planet. But in science, extraordinary claims require extraordinary proof. For the practicing scientist making such claims, a healthy dose of skepticism is essential, because someone else is sure to check your results very, very carefully.

See lacked that internal skepticism, and he would pay a steep price for its absence. In May 1899, a former student of See's named Forest Ray Moulton published a paper in the same *Astronomical Journal*, demonstrating that See's planet around 70 Ophiuchi couldn't exist because the laws of physics would not allow it.

Science is a "call and response" kind of business. Much as blues or jazz musicians will pick up on a riff that's played by one of their bandmates, See could have taken Moulton's results and built on them. He could have conceded that, with cutting-edge observations such as his, there were bound to be misinterpretations. He could have learned from the episode and built better science.

Instead, he doubled down.

In a blistering letter to the *Astronomical Journal*, See attacked Moulton and tried to weasel his way out of mistaken claims about planets. He wrote that he already knew about Moulton's objections and then waffled about the nature of the orbit and the planet. The editors of the journal were so taken aback by the acid tone of See's letter that they took the extraordinary step of printing only a few pieces of its text. Then they handed See the Victorian version of a smackdown: "The present is as fitting an opportunity as any to observe that heretofore Dr. See has been permitted, in the presentation of his views in this journal, the widest latitude that even a forced interpretation of the rules of catholicity would allow; but that hereafter he must not be surprised if these rules, whether as to soundness, pertinency, discreetness or propriety, are construed within what may appear to him unduly restricted limits."

The *Astronomical Journal* was essentially threatening See with censure.

Things went downhill from there. See's resentments and temperament led him from the world's greatest centers of astronomy down to the "Naval Observatory" at Mare Island, California. This was little more than a timekeeping station attached to a huge naval shipyard. Mare Island had no telescope worth mentioning.

Lacking access to a good instrument for observations, See

turned his attention to theory. Unfortunately, while he had clear talents with a telescope, his instincts for fundamental physics were terrible. See managed to miss the boat on every major revolution happening in physics at the turn of the century. He consistently rejected the profound discoveries about atomic phenomena in the new science of quantum physics, and he opposed Einstein's triumphant theory of relativity, claiming that his own ideas about cosmic structure had been proven by observation. (They had not.)

The final nail in the coffin of See's scientific reputation was a 1913 book called *The Unparalleled Discoveries of T.J.J. See*. The author called See "the greatest astronomer in the world." Upon further investigation, however, some suggested that it was See himself who'd written the book. He would never regain the respect of his peers, and he died in 1962, rejected by his chosen profession.

THE PROBLEM OF PRECISION

See would not be the last astronomer whose claims of a planet discovery would prove tenuous or career-threatening. A number of times in the years that followed, astronomers claimed to have detected a planet, only to see their claims evaporate. The difficulty in finding exoplanets can be summarized in a single word: *precision*. Planets are small, and stars are big. Planets are dim, and stars are bright. Planets are cold, and stars are hot. Planets have small masses, but stars weigh in as behemoths. The Sun, for example, would appear a trillion times brighter than the Earth when seen from the stars. That means trying to see an earthlike planet across interstellar distances would be like looking from New York City to

AT&T Park in San Francisco, where the Giants play, and making out a firefly next to one of the stadium spotlights.

So for scientists to "see" a distant planet, they must pull the tiny signal it produces out of the enormous impact of its star. There are a number of strategies astronomers can pursue to detect an exoplanet, but all demand high-precision measurements.

The basis for the oldest methods of detecting planets is the astrometry T.J.J. See was using, which focuses on the orbital motion of the star and planet. We usually think that planets orbit around their stars. The truth, however, is more interesting: objects always orbit *each other*. Binary stars of equal mass both circle around a point halfway between them. But if the mass of one of the objects is less than that of the other—as the case would be when a tiny planet orbits a big star—the orbit's center will be nearer to the center of the heavier object. So even though it *looks* like a planet orbits its star, the planet's gravity is still forcing the star to shuffle around in a tiny orbit. The center of that little dance is just slightly displaced from the star's own center.

See's astrometric studies were designed to see that tiny stellar motion. The idea was to track the position of a star over many years. In this way, astronomers would see the star zigzag as it was "perturbed" by the gravity of its unseen planet. But changes in the star's position as it wobbled back and forth would be minuscule. For example, aliens looking at the Sun from fifteen light-years away would have to strain to see the orbital wobble caused by even the most massive planet in our solar system. The precision needed to measure these tiny shifts in position was beyond the technology See had at his disposal.

There is another way of tracking the gravitational dance of a star and its planet, one that relies on tracking changes in the star's velocity rather than its position. As the star executes its little orbit, the gravity of the planet will cause it to swing first toward observers on Earth and then away. If astronomers could detect these changes in velocity—called *reflex motion*—it would constitute a detection of the orbiting planet. But like the orbits themselves, the changes in stellar velocity caused by orbital reflex motions are so small that taking measurements at the needed level of precision presented a huge technical challenge.

A third way of seeing an exoplanet focuses only on a star's brightness, meaning its total light output. During any given year, between two and five solar eclipses are visible from the Earth's surface. Each occurs when the Moon lines up just right for earth-bound observers, passing in front of the Sun and either partially or totally blocking its light. The same principle can be applied to planet hunting.

Imagine a distant star that hosts an exoplanet. Now imagine that the planet's orbit around its parent lines up perfectly with the "line of sight" between Earth and the star. That kind of alignment means the exoplanet will briefly swing between Earth and the star once during each of its orbits, just as the Moon swings between Earth and the Sun during an eclipse. Each time the planet gets between us and its star, it will block a fraction of the star's light, and from Earth we will see the star dim ever so slightly.

Astronomers use the term *transit* to describe a planet crossing the face of a star. Seeing an exoplanet transit its own star would require hyper-precise light detectors. Aliens looking at the Sun from interstellar distances would see its light dim by just one percent when Jupiter crossed its face. An Earth transit would dim the

Sun by just 0.01 percent. Along with this demand for precision, there is another complication. Stars can naturally produce light variations of the same order as an exoplanet transit. Dark regions on stars, called "spots," caused by powerful stellar magnetic fields, are just one of many sources of natural variation. Any successful transit-based exoplanet-hunting method would have to be exact in both its measurements and its understanding of the star being measured.

By the early 1970s, planets had been hiding beneath their veils of imprecision for so long that many scientists had given up on trying to find them. In addition, throughout the 1950s and 1960s, there had been enormous progress in other arenas of astronomy, like the study of distant galaxies. Hunting for other worlds came to seem like a dead end.

"I remember how, in the early 1990s, people would look down at the few researchers who were pushing for planet hunting," recalls one scientist. "There were NASA administrators who'd walk the other way just to avoid being bugged by them. It was a hard time for those guys."[2]

But the fortunes of the planet quest were about to change. The first steps toward taking exoplanets seriously began in the mid-1970s, and the motivation came directly from Frank Drake's original questions about the search for alien intelligence.

THE PATHS TO AN ANSWER

Frank Drake and Carl Sagan's very public discussions about exo-civilizations established the scientific basis for the search for extraterrestrial intelligence, or SETI. But the search itself would require a new generation of scientists. Chief among their number was Jill Tarter.

Like Drake, Tarter began her scientific training at Cornell, in the engineering physics program. But by the time she completed graduate school at the University of California, Berkeley, she'd decided to focus her work on SETI.[3] Over a long and distinguished career, Tarter carried out observational programs at radio observatories across the world, served as project scientist for NASA's SETI program, and was given the Bernard M. Oliver Chair at the SETI Institute.[4] She has seen firsthand how the question of exo-civilizations and the question of exoplanets converged.

In the 1970s, Tarter's dedication to SETI took her to a series of meetings where questions of precision and planet detection were first taken on in earnest. "Technology for finding planets just didn't exist back in the early 1970s," she says. "That means astronomers needed to get together and figure out exactly what the barriers were and how we could beat them."[5] With this goal in mind, in 1975 a workshop was organized at NASA's Ames Research Center in San Jose, at which the general problem of SETI technologies was first laid out. This workshop focused on search strategies for signals from exo-civilizations, but the attendees agreed that the factors in the Drake equation needed to be explored on their own as well. The most important of these sub-questions was the fraction of stars with planets and the fraction of planets in the habitable zone.[6]

"The original workshop led to two others that focused explicitly on planet-hunting methods," Tarter told me in an interview. "There was a meeting at [Ames] in 1978. This was the first time the different methods of planet hunting were drilled down into to see which one had the highest chance of success."

Records from that meeting show that most of the discussion focused on astrometric sky mapping, the approach that See used. Searches based on detecting reflex motions were discussed in detail, too. Direct detection—actually seeing the light from a planet—was also on the table.[7] But the transit method, based on the dimming of starlight due to a passing planet, didn't even make it into the report. The future would show the irony of this exclusion.

Though the problems with all the methods were acknowledged to be vast, the report ended on a positive note. "The prospects of increasing our confidence concerning the frequency and distribution of other planetary systems are good," the authors concluded.[8] Later, another SETI-inspired NASA workshop was held at the University of Maryland to explore technical details in more detail.

"People came away from that [second] meeting with a sense of what was possible," Tarter told me. "The reflex motion approach was

Astronomer and SETI research leader Jill Tarter.

seen as particularly promising if the technology could be hammered out. I think a lot of folks were really excited."[9]

Not everyone was so happy, however. While the transit method was raised at the Maryland meeting, its prospects were deemed to be dim. The final report concluded, "The Workshop considered the role of photometric [transit-based] studies in an effort to detect other planetary systems and upheld the conclusions of earlier studies, namely, that photometric studies are not practical."[10]

That conclusion didn't go down well with one tenacious scientist. "There was a young NASA researcher named Bill Borucki," Tarter said. "He felt the transit method had a lot of promise, even if everyone else thought it was hopeless. I think he was determined to prove them wrong."

THE FALL OF A THREE-THOUSAND-YEAR-OLD QUESTION

In 1995, at an astronomy conference in Florence, Swiss scientist Michel Mayor walked up through the audience and took his place at the podium. The other astronomers present looked around the room and wondered why a film crew had just appeared. That's when Mayor dropped his epoch-making bombshell. He and his partner Didier Queloz had firm evidence for the existence of another planet orbiting another star.[11] When it came to solar systems, at least, we were not alone.

In the decade and a half following the Ames and Maryland meetings, the hurdles blocking the way to reflex motion–based planet searches had been overcome. In the US, astronomers Geoff

Marcy and Paul Butler had built a series of ever more sensitive instruments to monitor a long list of nearby stars. Theirs was the world's most complete planet-hunting program.

But Marcy and Butler were expecting other solar systems to look like ours. They thought they'd need years of tracking before the signal of a Jupiter-sized planet in a Jupiter-sized orbit would appear in their data (Jupiter takes twelve years to make one swing around the Sun). The European researchers Mayor and Queloz had an observational program oriented toward finding close binary stars. They'd gotten lucky with their planet detection, but also had the insight to recognize what they'd found.[12]

Mayor and Queloz discovered their planet orbiting the star 51 Pegasi, which is fifty light-years from Earth (one light-year equals six trillion miles). The planet, called 51 Pegasi b, was Jupiter-sized, but swung around its star once every four days. That meant it was almost ten times closer to its star than our innermost planet, Mercury, is to the Sun.[13] A giant planet on a tiny orbit was not what astronomers were expecting when it came to solar systems.

Back in the US, Marcy and Butler quickly began looking for planets on such short orbits. It didn't take long for results to appear. At a press conference just a few months after Mayor's talk in Florence, Marcy and Butler announced their own discovery of two more Jupiter-sized worlds.[14]

New planets began to pile up after the discovery of 51 Pegasi b. As the shock of discovering exoplanets wore off, astronomers got down to the work of building a census of the new worlds.

But the real prize still lay in the search for Earth-sized worlds living in the star's habitable zone, where water and perhaps life could exist on the surface. The Earth's mass is one three-hundredth that of the Sun, a fact that meant even more precision was needed to detect Earth-sized worlds. That need for greater precision was also coupled with the problem that reflex motions only worked on one star at a time. What astronomers desperate for data needed was a precise way to discover planets wholesale. That threshold would be broken by the stubborn resilience of a man who simply refused to accept rejection.

Bill Borucki was a longtime NASA scientist who had cut his teeth on the physics of spacecraft heat shields. In the late 1970s, he decided to switch fields. The problem of planet detection offered the kind of technical challenges he loved, and after the Maryland meeting where transit-based planet-hunting methods had been dismissed, Borucki became determined to show these methods could work. In a now-famous 1984 paper, he and a coauthor laid out the basic framework for how to build a precise device to detect tiny changes in a star's light output. Then, in 1992, he proposed a space-based telescope using the same technology for planet hunting.[15]

While NASA thought the idea was interesting, it didn't believe Borucki's detectors would work. Unperturbed by the proposal's failure, Borucki began systematically addressing NASA's concerns. He built prototypes on the cheap to demonstrate that his system could hit the needed goals. After months of exhaustive work, Borucki's designs worked exactly as he said they would. In 1994, he spent months putting together the documentation needed to propose his transit-based telescope again. The proposal was rejected a second time.

A different set of concerns was raised in the second rejection. The new questions focused on Borucki's claim that he could do transit-detection on many stars at once. Once again, Borucki cobbled funds together and carried forward the extraordinary efforts needed to address each and every objection. Four years later, he and his team sent in their new version of the proposal. The proposal was shot down a third time.[16]

A reasonable person might have given up at that point. But in this regard, Borucki was not reasonable. He knew he was right. He knew the transit method would be a game changer. The only direction he could allow himself to move was forward.

Eventually, Borucki prevailed. After more than two decades of working on the same idea and having that idea rejected as scientifically unsound, Borucki's proposal was finally accepted. What would come to be known as the Kepler Mission was given the green light.[17]

Kepler was designed to stare at a single portion of the sky. In that small patch of cosmic real estate, about 156,000 individual stars had been identified as worthy of attention.[18] The satellite would patiently watch the same stars week after week, year after year. The patience was needed to accumulate enough transits—enough dips in light output—to provide an unambiguous signal of an orbiting exoplanet.

On March 6, 2009, the Kepler telescope rode into space on a Delta II rocket.[19] The launch was flawless. After so many years of rejection, Borucki and his team were staring across the frontier, ready to see how well his decade-spanning vision would work. They didn't have to wait long.

"As soon as the data started coming in from the spacecraft, we could see transits," recalls Natalie Batalha, a NASA astronomer who joined Borucki ten years earlier. "You could see the dips as clear as day. We were literally just sitting there in our office, watching as new planets were discovered with each transit."[20]

The first confirmed detections of exoplanets by Kepler came in January 2010, but they weren't the real news. Along with these detections, thousands of Kepler "candidates" were identified. These were stars showing dips in light that hadn't yet been confirmed as real planet detections. With so many exoplanet candidates, the Kepler team was sitting on the equivalent of a cosmic piñata. By 2014, that piñata had been busted wide open. That year, the Kepler team announced the discovery of 715 exoplanets in a single news release.[21] Wholesale planet hunting was the new reality. By 2015, the combination of Kepler and other methods had given astronomers 1,800 new worlds that were ready for detailed investigation.[22]

As the list of exoplanets grew, one of the first and most important conclusions was how different the architectures of other solar systems could be from our own.

Here on Earth, we all grew up learning about our solar system's tidy arrangement of small, rocky worlds tucked close to the Sun and larger gas giants splayed out at ever-greater orbital distances. The very first exoplanet discovered, 51 Pegasi b, showed that this arrangement was anything but universal. It's an example of what is called a "hot Jupiter"—a gas giant that somehow ended up on an outrageously tight orbit. Big planets on small orbits are easy to find in reflex-motion searchers, so many more of these hot Jupiters were quickly added to the exoplanet tally. Lots of stars were also found to have Jupiter-sized worlds on orbits the size of Earth's, rather than out at the farther reaches of their solar systems.

Eventually, other kinds of planets living close to their parent stars would be found—"hot Neptunes" and even "hot Earths." Inner rocky worlds and outer gas giants were clearly not the only way nature laid out her planetary families. Systems with hot Jupiters were the most dramatic examples of "weird" solar systems, but there were many other surprises. Systems consisting of *only* smaller rocky worlds were found, and even they looked weird by our standards.

"One of the big surprises was our discovery of what we call 'compact multis,'" says Batalha. "These are planetary systems with a bunch of small planets clustered very close to each other."[23] In our solar system, Earth and Venus are the nearest neighbors, coming as close to each other as twenty-five million miles. That's why it takes many months for us to reach these worlds. But in the compact multiplanet system Kepler 42, for example, there are three planets stuffed into remarkably tight orbits. These worlds get one hundred times as close to each other as Venus ever gets to Earth.[24] If you lived on one of Kepler 42's worlds, you could travel to your neighbor planet in just a week or so, using the kind of spacecraft that got us to the Moon back in 1969.

The architecture of planetary systems wasn't the only surprise. "We found a whole class of planet out there that don't even occur in our solar system," says Batalha. There are no planets orbiting our Sun with a mass between those of Earth and Neptune. That represents a considerable gap since Neptune is a big mix of gas and ice and weighs in at fourteen times the mass of Earth. Earth and Neptune are, in other words, very different kinds of planets. But as the exoplanet revolution matured, astronomers soon found worlds—a lot of them—with masses right in that gap between one and fourteen Earth masses. They called these "super-Earths," and it soon became clear that this new kind of planet, which doesn't even occur in our solar system, might be the most common in the universe.[25]

"We don't even understand what these worlds will look like," says Batalha. "Some of them could be rocky. But some could be water worlds with deep oceans surrounded by thick water-vapor atmospheres. Others could be a mix of rock and ice and gas. The possibilities are pretty broad."

Beyond the general findings, there were the incredibly weird specific cases. For example, there's J1407B, the "super-Saturn" located 434 light-years from Earth. The rings orbiting this gas giant stretch two hundred times farther than the gossamer disk surrounding Saturn.[26] Then there's 55 Cancri e, which is forty light-years away. Its diameter is only twice as great as Earth's, but it has a mass almost eight times higher, resulting in a density so great that it may be a planet made of diamond.[27] And not to be missed is the ominously named WASP-12b. It's a hot Jupiter with a temperature of nearly 4,100 degrees Fahrenheit, making it one of the hottest exoplanets ever discovered. Astronomers can see a trail of debris surrounding the planet as WASP-12b boils away in a torrent of evaporating gas.[28]

In the end, though, what matters most are not hot Jupiters, super-Saturns, or super-Earths. The numbers as a whole are what make the exoplanet revolution so important for us. At the beginning of the second decade of the second millennium of the Common Era, humanity finally learned that, in one very real sense, we were not alone. There were other worlds out there. Just as important, with a full census of planets being built, the first three terms in the Drake equation were now fully known. With that advance, questions not only about planets, but even about civilizations other than our own, could be seen in an entirely new light.

DRAKE AND THE EXOPLANET REVOLUTION

The first term in Drake's equation describes the rate of making stars (called N_*). It has been known with some accuracy since the late 1950s, and subsequent studies have only refined that value (about one star per year).[29] But when Drake first wrote his equation in 1961, the second term, describing the fraction of stars with planets (called f_p), and the third term, describing the number of planets in a star's habitable zone (called n_p), were anyone's guess. By 2014, in the wake of Kepler and other exoplanet studies, there was enough data in hand to give scientists meaningful—that is, statistically significant— values for those numbers.

The implications of this advance are stunning enough to change our experience of the night sky. Let's consider the fraction of stars with planets first. Remember that, during the early part of the twentieth century, astronomers believed planet formation was a rare event, meaning the fraction of stars with planets would be very low. But by 2014, the agreed-upon value for f_p was about 1.[30] In other words, pretty much every star you see in the night sky hosts at least one planet.

The next time you find yourself outside at night, take a moment to stop and consider the implications of this result as you gaze at all those pinpricks of light. Every one of them hosts at least one world, and most stars will have more than one planet. Solar systems are the rule and not the exception. They're everywhere.

The advent of Kepler also allowed astronomers to reach a firm conclusion about the average number of habitable-zone planets orbiting each star. Remember that the habitable or Goldilocks zone is a band of orbits around a star where liquid water can exist on a planet's surface. That means any planet in a star's habitable zone might be a world of rain and rivers and oceans—a world potentially capable of supporting life. There are currently two planets in the Sun's habitable zone—Earth and Mars—and both have had water running in torrents across their surfaces.

From the exoplanet data, astronomers can now say with confidence that one out of every five stars hosts a world where life as we know it could form.[31] So, when you're standing out there under the night sky, choose five random stars. Chances are, one of them has a world in its Goldilocks zone where liquid water could be flowing across its surface and life might already exist.

The importance of the achievement represented by nailing these two numbers cannot be overstated. Through the hard-won efforts of a generation of astronomers, we increased the number of known terms in Drake's equation by 200 percent. Where there was darkness, there now is light. Where there was ignorance, there now is knowledge.

YES, THERE PROBABLY HAVE BEEN ALIENS

But what, if anything, could the trove of data leading us to these numbers reveal about the possibility of other worlds inhabited by technology-deploying, civilization-building species? We still have zero evidence that such civilizations exist. Is there any way to leverage the achievement of the exoplanet revolution to say something—*anything*—about exo-civilizations? Addressing exactly that question was the task Woody Sullivan and I took on at the beginning of 2015.

I first met Woody Sullivan in the late 1980s, when I was a physics graduate student at the University of Washington. He's tall and slender with a wry sense of humor and a passion for sundials and baseball (the Seattle Mariners, in particular). Most importantly, Woody is a radio astronomer with an unwavering interest in SETI. When I was a graduate student, he was the only person on the faculty at the University of Washington who worked on the question of exo-civilizations. This was well before NASA began serious funding for astrobiology. The exoplanet revolution was still a decade from its inception. In the 1980s, SETI and its astrobiological surroundings were still considered a bit "out there" for many folks. But Woody didn't care. He was interested, and he thought there was science to be done. So he pressed on and wrote a number of important papers on the subject.

I once helped Woody teach a course called "Life in the Universe." He set the class up to deal with everything from the nature of physical law to the prospects for life on other worlds. His perspective was broad and imaginative. I loved being involved with that course, and its perspectives shaped my thinking for decades. It was also the first time Woody and I started talking about exo-civilizations. Those conversations have been going on ever since, even before I did any direct work in astrobiology.

In 2014, Woody and I found ourselves asking if all the new exoplanet data could be used to infer a definite conclusion about technological civilizations on other worlds. The astonishing progress made since the first exoplanet discovery had to be good for something. Wasn't there some way to use it with an eye toward answering Drake's original question about our uniqueness in the universe? We soon saw there was a path forward, but to take it, we'd have to turn Drake on his head.

Drake built his famous equation on a simple question: How many exo-civilizations exist now? He chose that focus because his real interest was in finding signals from alien civilizations. For his equation to make sense, those aliens had to be out there, emitting radio signals right now (relatively speaking). But to make the kind of progress Woody and I were interested in, we realized we had to change the focus. We had to ask a different question—one that could be answered by the exoplanet data. Our new question was only slightly different, but the small change we made would mean everything in terms of results. Our question was this: How many exo-civilizations have there ever been across the entire history of the universe?

Taking this approach gave us a strategy for getting an empirically based number concerning the existence of exo-civilizations. First, we combined all the astronomical terms in Drake's equation into one. This was easy, since they were all known. Then we began thinking differently about three unknown probabilities involving life in Drake's equation (f_l, f_i, and f_t). Rather than dealing with them separately, our approach lumped them all together, too. We were interested in the process as a whole, going from the origin of life all the way up to an advanced civilization. We called our new term the "bio-technical probability," f_{bt}, and it is the product of multiplying all the usual life-centric terms in the Drake equation together. In the language of math:

$$f_{bt} = f_l f_i f_c$$

Finally, by asking about the total number of exo-civilizations that had *ever* existed, rather than limiting our interest to those existing now, we took the issue of the average lifespan of a civilization out of the problem. We didn't care if the exo-civilization overlapped with our own. It didn't matter. We just cared that they had existed at some point in cosmic history. Effectively, that allowed us to ignore the final factor—the pesky lifetime term, L—in Drake's equation.

Our approach gave us a new form of Drake's equation that looked a lot simpler:

$$A = f_a f_{bt}$$

In this version of the equation, A was just the total number of civilizations that had ever existed. We thought of A as standing for "archaeology" because, in a weird way, that's what we were interested in. Because we took the whole of cosmic history as our playing field, most of the civilizations we'd be describing in our approach would probably be long gone. But all that mattered to us was that they had existed at some point in cosmic space and time. That was the archeological bent our approach took. We saw that the Kepler data could tell us more about what *had* happened than what was happening right now.

Meanwhile, f_a represented all the astronomical terms in the original equation. The important point was that, since all of those terms were now known, f_a was also known. That left just the biotechnical probability (f_{bt}). It represented all the unknown, life-oriented probabilities in Drake's equation. This was what we were after.

By rewriting the equation without L and using the new exo-planet data, we then saw that we could recast the question of the probability of alien life in a way that turned our new form of the equation into a very specific and scientifically meaningful formulation. Our new question, therefore, was: What would the biotechnical probability per habitable zone planet have to be for humans to be the only civilization nature had ever produced over the entire history of the universe?

In other words, what were the chances that ours is the only civilization ever? Putting in the exoplanet data, we found the answer to be 10^{-22}, or one in ten billion trillion.[32] We called this number the "pessimism line," for reasons we'll unpack below. To me, the implications of this number are staggering.

To understand how to think about the pessimism line, imagine you were handed a very big bag of Goldilocks-zone planets. Our results say the only way human beings are unique as a civilization-building species would be if you pulled out ten billion trillion planets and not one of them had a civilization. That's because Kepler has shown us that there must be ten billion trillion Goldilocks-zone planets in the universe. So the pessimism line is really telling us how bad the probability of a civilization forming would have to be in order for ours to be only one that has ever existed.

Ten billion trillion planets is a lot of worlds to go through without finding anything. The sheer size of that number is enough to make it seem like we are not the first time nature has ever created a civilization-building species. By comparison, think about getting killed by lightning, an event most of us think of as unlikely. The probability that you'll be killed by a lightning strike in any given year is about one in ten million. But, based on the pessimism line, your lightning-induced death is a thousand trillion times more likely than humanity being the only civilization in cosmic history. Surely nature is not that biased against evolving civilizations? It can't be that perverse. Or can it?

Drake's question—How many civilizations exist now?—still can't be answered. But our question—What limit can be placed on the odds that it's ever happened?—could be. We could put a stake in the ground and say that if nature's processes of evolution led to odds less than the pessimism line, then yes, ours is the only energy-intensive, technological civilization that's ever existed. But if nature's value for the biotechnical probability is higher than one in ten billion trillion, then we are not the first.

After our paper was published in the journal *Astrobiology*, I wrote an op-ed for the *New York Times* about our result. The *Times* ran the piece with the headline "Yes, There Have Been Aliens." Within days, I was inundated with requests for interviews from outlets ranging from the large and established, like CBS, to small websites run by avid UFOlogists. Some of those folks might have been discouraged from contacting me if the headline had been closer to what we really meant, which was "Yes, Aliens Probably Existed." But either way, our result was bound to generate controversy. The critiques are worth looking at closely, since interpreting the pessimism line correctly is critical.

Our goal, after all, is to see how astrobiology and the study of life on other planets can help us understand climate change and the project of civilization on our own world. In that pursuit, the pessimism line marks a critical boundary where we might see our project set against the stars. But to truly understand what the pessimism line can do for us in that endeavor, we must first understand what it cannot.

THE CRITIQUE

One of the principal objections raised to our paper (and the *New York Times* op-ed) was straightforward. Just because the probability that we're the only civilization in cosmic history is low (10^{-22}), that doesn't constitute a proof that exo-civilizations have existed before us. This was the argument made by Ross Andersen, the science editor for the *Atlantic*, and Ethan Siegel, an astrophysicist who writes for *Forbes*.[33] Andersen and Siegel are excellent thinkers, and their criticisms contained a lot of insight. Their essays cut to the heart of key issues in what Woody and I were trying to explore. Most of all, their skepticism made me think even harder about the ideas in our paper, and I was grateful for that.

There was one point in particular that Andersen took issue with, and it was embodied in this sentence from my *Times* op-ed: "The degree of pessimism required to doubt the existence, at some point in time, of an advanced extraterrestrial civilization borders on the irrational."[34] He was right to criticize that line. In spite of the bar set by the pessimism line, it's not "irrational" to think we are unique in cosmic history. In fact, the only empirically valid claim Woody and I can make is this: we can say with certainty where the pessimism line lies. In the absence of more data, it is rationally possible to construct an argument that nature's value for the biotechnical probability lies below 10^{-22}.

Questions were also raised about the values for the individual pieces that make up our biotechnical probability. Some argued that the probability of making just simple forms of life would be too low to allow civilizations to ever form. Or perhaps it was the probability of life evolving its way up to intelligence that was really low. But these considerations don't change our result. Our biotechnical probability, f_{bt}, does not hide the fact that each of the life-centric terms in the Drake equation might be small on its own. We didn't establish our pessimism line by ignoring possibly small values for the individual life-centric terms. Instead, our reworking of the Drake equation bundled them all together. Our approach let us go for the whole enchilada at once: the entire evolutionary process, from abiogenesis up to the creation of a technological civilization. No matter how improbable you think each individual step is, it's the total probability of other civilizations existing that matters. That's what you have to pay attention to, and that's what the pessimism line represents.

We called our result the pessimism line for good reason. The whole history of the debate about life beyond Earth is an argument between optimists and pessimists. It's a debate that began with the opposition between Aristotle and Epicurus, extended through the 1800s to Flammarion versus Whewell, and took its modern turn with the Drake equation, through which the battle between pessimism and optimism became quantitative.

Since the 1961 Green Bank meeting, many scientists have argued that exo-civilizations are rare. What is rarely specified, however, is exactly what "rare" really means. Scratch below the surface, and you'll see that many self-described pessimists' version of rare is way above our pessimism line. That's why the history of the debate can't be ignored.

Looking across that debate since the Drake equation appeared, we see that optimism always has a clear upper limit. You can't get more optimistic about the possibility of life evolving on an exoplanet than if you say it *always* occurs (that would mean setting the value of f_l at 1). The same holds true for the other life-centric terms in the Drake equation. You can't assign a value greater than 1 to the probability of intelligence—or high technology—evolving. Making all these choices implies that every exoplanet in the habitable zone will create life that goes on to form an intelligent technological civilization.

But pessimism is another story. How low is low? How pessimistic do you have to be—expressed in terms of the Drake equation—to be truly pessimistic about exo-civilizations? That was what Woody and I were asking. Our answer provided a line marking the limit of true pessimism. If nature had a biotechnical probability that was lower than our limit—one in ten billion trillion—then human beings had to be the only example of a high-tech civilization in the history of the observable universe. In that case, we'd be truly and deeply alone in the most absolutely cosmic sense of the word. But if the forces of evolution led to a number higher than the pessimism line, then what's happened with us on Earth has happened before.

Of course, we still don't know what nature has chosen. But to see what our next steps might be in thinking about exo-civilizations and our own fate, we can look at how our pessimism line compares with what actual pessimists have proposed for the biotechnical probability.

Pessimist #1: Ernst Mayr. Chief among the exo-civilization pessimists was the renowned German evolutionary biologist Ernst Mayr.

Mayr was a brilliant scholar who was instrumental in linking classical ideas from Darwin to the revolution in genetics that occurred after the discovery of DNA. But Mayr never bought Carl Sagan's optimism about SETI or the existence of other intelligent forms of life. In 1995, the Planetary Society gave both men the chance to voice their opinions on the subject and respond to each other's criticism. While Mayr never provided an explicit value for the biotechnical probability, from his essay[35] we can extract an estimate of his pessimism.

Mayr had no doubts about life forming on other planets. Of the probability that life exists elsewhere in the universe, he wrote, "Even most skeptics of the SETI project will answer this question optimistically." Because molecules necessary for the formation of life had been found in cosmic dust, he conceded that it was very possible there was life elsewhere.

The development of intelligence, however, is where Mayr's pessimism kicks in. Looking at the history of Earth, Mayr wrote, "Only one of these [approximately fifty billion species that have existed on Earth] achieved the kind of intelligence needed to establish a civilization." And on the subject of intelligence leading to a civilization, Mayr wrote, "Only one of [the twenty or more civilizations that have risen in the past ten thousand years] . . . reached a level of technology that has enabled them to send signals into space and to receive them."

From Mayr's statements, we can estimate what he thinks the biotechnical probability might be. Given that he argues that the formation of life is not a hard step, let's assume he would be happy with a value of one in a hundred for that factor (10^{-2}). After all, something that happens once every hundred times is not really very rare.

Given his statement about the total number of species evolved on Earth versus the single one that became intelligent, we can infer that Mayr might say that the odds of evolving intelligence on any given exoplanet with simple life would be one in fifty billion (or about 10^{-11}). That certainly seems pretty pessimistic. Finally, from his statements about civilizations becoming high tech, we might infer he'd consider the probability for that term to be one in twenty. Let's err on the side of pessimism and call this one in a hundred (10^{-2}).

If we put all of these together, we would find that Mayr seems to be arguing that the value of the biotechnical probability is around one in a thousand trillion (10^{-15}). That is certainly pretty small. Recall that if Mayr is right, you would have to sort through a bag of one thousand trillion planets to find a single technological civilization. Given that there are "only" one hundred billion stars in our galaxy, Mayr's brand of pessimism would mean we were alone in our galaxy.

But being alone in the galaxy and being the only civilization the universe has ever produced are two different things. Comparing Mayr's pessimism with the limit expressed by the pessimism line Woody and I derived shows something remarkable.

Even if civilizations were as rare as Mayr proposes, there is still a vast gulf between Mayr's "one in a thousand trillion" and the pessimism line's "one in ten billion trillion." To be exact, even if Mayr is correct, there will still have been ten million high-tech civilizations appearing across space and time. That means ten million individual stories of a species waking up to itself. Ten million different versions of science being harnessed to harvest a planet's resources and build a civilization. Ten million different histories of civilizations either going on to become long-lasting or collapsing under the weight of their own choices.

If you tried to imagine the history of each of these civilizations, giving each one an hour of your time, it would take 1,140 years to get through them all. That's how many exo-civilizations would have existed in what Mayr thought to be a pessimistic universe.

Pessimist #2: Brandon Carter. In 1983, the physicist Brandon Carter developed an absolutely ingenious argument against exo-civilizations. Carter was famous for using simple observations to infer immensely vast conclusions about the universe and our place in it.

His thinking about exo-civilizations began with the simple observation that the time required for intelligence to arise on Earth was close to the total age of the Sun. In particular, while the Earth has been habitable for four billion years, it will only remain so for another billion or so years because the Sun is continually heating up. It will eventually grow so hot that the Earth's orbit will no longer be in the habitable zone. Thus, a technological civilization (ours) has only appeared on Earth close to the end of its period of habitability. Using this one fact, Carter made the case that intelligence must have required evolution to pass through a series of "hard steps." Fulfilling each of these hard steps would itself be highly improbable.[36]

Looking at Earth's evolutionary history, Carter argued that there were ten evolutionary hard steps. These included the evolution of oxygenic photosynthesis or of multicelled animals. Based on these ten hard steps, he devised a calculation to predict the probability of exo-civilizations, which was just our biotechnical probability by a different name. Carter's value came out to be 10^{-20}. He claimed this was "more than sufficient to ensure that our stage of development is unique in the visible universe."

What is wonderful about Carter's calculation is that it leads to an explicit number for the biotechnical probability. The number he calculated was so small, it implied to him that no technological civilization other than our own could ever have existed across all cosmic space and time.

But that's not what Carter's number implies! A comparison of the pessimism line Woody and I found with Carter's result shows that his 1983 calculation still allows for one hundred exo-civilizations. Carter intended his calculation to be hyper-pessimistic, but it turns out to be optimistic instead. Carter's original argument still leaves us with the remarkable conclusion that we are not the first. If Carter is correct, a hundred other civilizations had passed through the processes of civilization building that we find ourselves in now.

It should also be noted that researchers who have followed Carter's line of reasoning now believe only five hard steps exist, if any exist at all.[37] This consideration, combined with the other values in Carter's original paper, implies a biotechnical probability of 10^{-10}. Compare that with our pessimism line, and you end up with a trillion exo-civilizations across cosmic history. Allowing for the existence of a trillion other civilizations is anything but pessimistic.

Pessimist #3: Hubert Yockey. Of course, one can find ways to argue for a hyper-*hyper*-pessimistic viewpoint. This is exactly what Hubert Yockey did in a 1977 paper. Yockey was a physicist and information theorist. His argument focused on the first life-oriented term in Drake's equation—the probability of life forming on an exoplanet. What are the odds, he asked, that random chemical combinations would produce the right kind of self-reproducing molecule for getting life started? His answer was less than an astonishing one in a trillion trillion trillion trillion trillion (his actual value was 10^{-65}).[38] This number is certainly below our pessimism line, and if Yockey is right, then we represent the only time in cosmic history that life of any kind has emerged.

But this kind of argument is balanced by the fact that there are strong counterarguments that life's emergence may not be so hard to achieve. Many of these responses come from advances in biology. For example, biologist Wentao Ma and collaborators used computer simulations to show that the first replicating molecules could have been short strands of RNA (a molecule closely related to DNA and an integral part of cellular machinery). These are much easier to form than what Yockey was thinking about.[39] Many researchers also take the fact that life appeared so quickly after the Earth's formation as an indication that abiogenesis may not be extremely hard to achieve. Either way, Yockey's hyper-hyper-pessimism seems to be an outlier in the debate about alien life.

THE BIG STEP

The pessimism line doesn't prove that other civilizations on other worlds ever existed. It doesn't help us in our search for signals from other civilizations that may overlap with our own. So what, exactly, does it allow us to say or to do?

More than anything, what Woody Sullivan and I did was use exoplanet science to raise a key philosophical point that drew its potency from real observations. It was an opening into a way of thinking about our place in the universe and the challenges of our Anthropocene moment in a radically different way from what we're doing now.

While the Drake equation was all about making contact with other civilizations, our perspective was straightforward: the exoplanet data now lets us make a reasonable argument that there have been many other civilizations before ours. If you agree that the pessimism line is low enough to make those other civilizations far more probable than we could have known before, then you can also take the step of considering them worthy of serious consideration. With that step, something remarkable can follow in facing the challenge of the Anthropocene.

Before we go any further, let me be clear that you don't have to make that step. Remaining deeply agnostic about the existence of other civilizations in cosmic history is certainly a stance that science cannot argue against. So, if you don't think it's worth considering those other civilizations seriously, that's fine. Everything we have already explored about astrobiology and the Anthropocene will still hold true. Our understanding of the climate change we are driving today must still be seen as grounded in what we learned by studying other planets in the solar system. And our questions about what to do next must still be informed by the understanding we gained by studying our own planet's long history of coevolution between the biosphere and Earth's other coupled systems. We know what we know because we have already learned a lot about what it means to think like a planet. That means there are rules to the game when it comes to the evolution of the Earth, including the Earth with us on it. That perspective alone undercuts the arguments of climate denialists and represents a fundamental shift in how we understand ourselves and the challenges before us.

But if you are willing to see the pessimism line as the universe's invitation to consider other civilizations seriously, then we can begin to ask what other civilizations mean for us. The purpose here is not to consider them as the source of science fiction stories, but to recognize that we are probably not the first experiment in civilization building the universe has run.

Throughout all of history, our mythologies have told us who we are, what we are, and where we stand in relation to the cosmos. But those stories ignored the possibility that we are one of many. Our stories did not—because they *could* not—include the possibility that our civilization was a planetary phenomenon that was not unique. That is why the exoplanet revolution and all the astrobiology we have explored so far can be a kind of wake-up call for us. It can be part of our coming of age as a civilization.

The discovery that the universe is teeming with habitable-zone planets connects the challenges we face in the Anthropocene directly with the questions Fermi, Drake, and Sagan asked fifty years ago. The pessimism line tells us that the universe has had lots of opportunities to do what it did on Earth. With that information, we can begin to seriously consider that there have been many other stories, meaning many other histories, beyond our own. It's an invitation to begin putting ourselves and our choices into a more accurate and fully cosmic context. If we take that step, then everything we've learned about planets and climate and biospheres becomes relevant to those other civilizations, too. We can treat those other civilizations as objects of study. That is why a science of those civilizations—a theoretical archaeology of exo-civilizations—is the territory we must explore next.

CHAPTER 5

THE FINAL FACTOR

MANY WORLDS, MANY FATES

Two decades into the twenty-first century, we find ourselves facing the existential challenge of creating a sustainable version of human civilization. The scale of human activities is pushing hard on the tightly linked planetary systems that make up Earth's climate. As the planet begins to move off into a different climate state, our project of civilization will, at the very least, find itself under stress. At worst, Earth's changes may make our project impossible to maintain.

We urgently need to adapt civilization so that it can continue for the long term, so that it can become fully and globally sustainable. But before we can start working toward that goal, there's an equally urgent question that often goes unstated: How do we know that's even possible? How do we know there *is* such a thing as a long-term version of our kind of civilization? Most discussions of the sustainability crisis focus on strategies for developing new forms of energy or the projected benefits of different socioeconomic policies. But because we're stuck looking at what's happening to us as a singular phenomenon—a one-time story—we don't think to step back and ask this kind of broader question. To even in pose it seems defeatist. But it must be addressed if we are to make the most informed, intelligent bets on the future.

Let's be clear about what our question implies. Maybe the universe just doesn't do long-term, sustainable versions of civilizations like ours. Maybe it's not something that's ever worked out, even across all the planets orbiting all the stars throughout all of space and time. Maybe every technological civilization like ours has been just a flash in the pan, lighting up the cosmos with its brilliance for a few centuries, or even a few millennia, before fading back to darkness.

This question speaks directly to Fermi's Paradox. Perhaps the bottleneck we face today explains the Great Silence of the stars. Our question points to the final factor in Drake's equation—the average lifetime of civilizations. Even if every planet orbiting every star in the universe evolved a civilization, it would still be possible that none lasted very long. That kind of fate might be universal for exactly the same reasons we find our own future challenged.

So, does anyone make it past the challenge we now face?

Staring down that question is where the rubber really meets the road in the astrobiology of the Anthropocene. The pessimism line tells us that, unless the universe is highly biased against the appearance and evolution of civilizations, others came before us. Each of those civilizations will have had a trajectory of development in terms of their growth and their impacts on their planets. Those trajectories are what we want to understand. Given what we have learned about planets and climate, there are good reasons to argue that many planets evolving a young, energy-intensive civilization will be driven into an Anthropocene-like transition. If there have been exo-civilizations before us, we've already learned enough about "thinking like a planet" to see if the conditions leading to Anthropocenes are common or rare. So, how can we use the science we know, gained from the planets we have seen, to begin a science of the civilizations we haven't?

HOW NOT TO DO A SCIENCE OF EXO-CIVILIZATIONS

Prosthetic foreheads. That's what you want to avoid—the Klingons, the Vulcans, the UFO aliens with the big heads. Science fiction has given us enduring images of alien races. Not surprisingly, most of them look a lot like us, but with different kinds of foreheads or ears or a different number of fingers on their hands.

In developing our science of exo-civilizations, we're not interested in what aliens might look like or how they might behave. We're going to avoid the specifics of their biology and their sociology because science provides us little to work with on those issues.

So, what issues can science help us with? There are three terms from Drake's equation that make up the biotechnical probability. They involve basic biology (the origin of life), evolutionary biology (the rise of intelligence), and sociology (the development of societies). When it comes to what might happen on other planets, we are on murky ground for each of these terms. But if we ask the right questions, there are principles that constrain our theoretical explorations. These constraints are like guide rails keeping our theoretical bowling balls from plunging into the gutter.

For the basics of life, for example, we are going to have to rely on our knowledge of chemistry. But we already know that chemistry works the same way in distant regions of the cosmos as it does on Earth. From observations of interstellar clouds, planet-forming disks, and even exoplanet atmospheres, we can see physics and chemistry playing out exactly as they do down here on Earth. So, no matter what surprises life on other worlds may have for us, it must still utilize the same basic laws of physics and chemistry that apply on Earth. Based on this cosmic uniformity, scientists are already exploring what alternative biochemistries might look like.[1] There are even studies of how photosynthesis might work on planets with very different kinds of suns.[2]

On the question of intelligence, things get shakier. That's because there are so many steps needed for its development. Worse, we don't know which steps are essential and which happened to be specific to how intelligence worked out on Earth. In dealing with the evolution of intelligence, however, we do at least have a principle we believe should be general across all planets. The genius of Darwinian evolution is its ubiquity. Darwin proposed that all life on Earth was shaped by the same set of simple processes: mutation, adaptation, and survival of the fittest. Simply put, whatever organism is best adapted to the environment will outlast its competition. It's a principle that applies to everything from the first self-replicating molecules to modern, fully-formed biological organisms. It should even apply to future self-replicating robots if we ever make them.

So, when it comes to evolution on other worlds, this kind of uniformity should prove useful, particularly when we think globally in terms of biospheres. Darwinian evolution, in terms of population growth and competition in ecosystems, gives us a constraint for our ideas as we follow them to their consequences.

The science of sociology and the question of the formation of civilizations seem to be another story entirely. We cannot assume that sociological truths we've observed on our world will hold true across time and space. Do other civilizations have political parties? Do they worship a god or gods? We can tell stories about how an exo-civilization might organize itself, but our descriptions would always be just that—a story. Here, I am specifically referring to questions of their morality or economics or religion. Have they, for example, created institutions that value altruism over conflict, or conflict over altruism? Does the idea of institutions even make sense in their civilizations?

Unlike the foundational laws of physics and chemistry or the potential for Darwinian evolution to be cosmically general, it's hard to see what kind of universal principles exist that would allow us to constrain something like alien economics. When it comes to sociology, I don't believe such constraints exist.

So, while we are now in a position to begin building a science of exo-civilizations, the questions we can meaningfully take on must be limited. We need to avoid science fiction stories. That means speculation about whether civilizations are warlike or peaceful, or whether cultures focus on empire building or are content to stay at home, is out of bounds. Trying to answer questions about any of these dichotomies is close to a hopeless task. Extending our knowledge from the seen to the unseen requires something that keeps our theory building within nature's possible bounds. No matter how far we want to reach, there has to be some ground for our feet to stand upon. For the time being, that means sticking with the physics and chemistry of planets (things like climate) and the parts of biology one can reasonably argue should be common. In developing our science of exo-civilizations, we should try to avoid questions about culture. That will be the challenge in building our astrobiology of the Anthropocene.

Of course, this strategy cuts out a whole lot of questions that many people want to know about exo-civilizations. For example: What are aliens like? Do they have two sexes, or twenty-three? Have they built a society on logic or on love? Are they traders or warriors? And of course: Do they look like us? If those are the questions you want to ask, I'm afraid you're out of luck as far as our scientifically bounded theorizing is concerned.

But there is one specific kind of question about those civilizations that our science of exo-civilizations can address directly. By sticking to the laws of planets we learned through Carl Sagan, Jack James, Lynn Margulis, James Lovelock, and thousands of others, we can now ask the question that matters most to our project of civilization: How common is the Anthropocene? How often do civilizations trigger climate change on their planets? And, most important, how easy is it for a civilization to make it through its Anthropocene bottleneck?

OF PREDATORS AND PREY

The Adriatic Sea has fed Italy's eastern shores for eight thousand years. From Venice in the north to Brindisi in the south, its warm

waters have provided a livelihood for more than a hundred generations of fishermen. There are 450 different species of fish swimming
in the Adriatic, many of which end up as food on Italian tables.[3] But
those tables have always been demanding. Human fishermen are the
Adriatic's top predator, and many of the sea's species are currently in
danger of collapse from overharvesting.

But the beat of fishermen's oars or the buzz of their motors in
the Adriatic has not been uniform across history. Conflict can slow
the pace of fishing, as fleets of warships patrolling coasts make the
work even more dangerous than normal. In World War I, the Adriatic became a battle zone. The new efficiency of mechanized navies
gave Italy's enemies a long enough reach that commercial fishing in
the Adriatic almost ground to a halt.

For all its hardship, that lull in fishing proved to be an unlikely
gift to science. By slowing the human draw on the Adriatic's fishing stocks, a paradox surfaced that reshaped how biologists thought
about animal populations, ecology, and the nature of their own work.

In the years immediately after the war, a young marine biologist named Umberto D'Ancona was working himself to exhaustion
studying fish populations and their evolution. Through long, diligent
work, D'Ancona amassed statistics on sales at fish markets in cities
like Trieste, Fiume, and Venice, across the length of Italy's Adriatic
coast. His data bracketed the war years, beginning with 1910 and
ending in 1923. Poring over the numbers, D'Ancona saw something
that defied explanation.

During the war years, when fishing had been reduced, the number of predators such as sharks seemed to soar. This might have
made sense if the numbers of prey fish, like mackerel, had also
climbed, as D'Ancona had expected. More prey should mean more

predators. But the numbers of prey fish didn't rise during the war. Instead, they dropped. The statistics in front of D'Ancona told him that less fishing led to fewer prey fish and more predators. The young scientist puzzled over his paradox until, in desperation, he brought his biology problem to an unlikely consultant: the great mathematician and physicist Vito Volterra.[4]

Volterra was a world leader in solving hard physics problems. His work had touched on everything from the structure of crystals to the behavior of fluids.[5] But Volterra's reputation was not the main reason fate and D'Ancona brought him into the domain of biology. D'Ancona was also marrying the professor's daughter. Luisa Volterra was herself a scientist, with a specialization in ecology—the biological study of populations and their environment.

Physicist Vito Volterra (third from left) developed the predator-prey model of population ecology for his son-in-law, marine biologist Umberto D'Ancona (far right). D'Ancona's wife, ecologist Luisa Volterra (daughter of Vito), stands next to him (circa 1930).

At the time Volterra took up the problem, mathematical "modeling" of the kind found in physics was not yet in the biologists' toolkit. Biologists certainly dealt with statistics, but modeling is something different. Modeling is essentially a theoretical enterprise. It's a process that begins by choosing a set of assumptions about how the world works. Those assumptions then get turned into mathematical equations, and those equations are what scientists mean what they talk about a model.

As we have seen in building climate models for Earth or Mars, the essential step in mathematical modeling is solving equations. Those solutions are descriptions of the world's behavior over time. They are, therefore, predictions. So, whatever equations Volterra came up with for D'Ancona's fish problem, their solutions needed to predict how the predator and prey populations changed with time.

Physicists have been making mathematical models ever since Newton devised his laws of mechanics back in the 1600s, giving physics its deeply theoretical emphasis. But biologists in the early twentieth century saw their work in a different way. The kind of modeling physicists routinely carried out didn't seem up to the task of explaining the complexity of living systems and their interactions. As complicated as the orbits of planets might be, the complexity of a single cell, or even a simple food chain, puts astronomers to shame. For biologists, fieldwork always led the way.[6]

By the time Volterra began thinking about his son-in-law's problem, however, things were changing. A movement had already begun to bring theory, in the form of mathematical models, into biology. That work had begun in the 1800s, when Pierre Verhulst of Brussels discovered what he claimed was a law of populations.[7] Consider, for example, a few bacteria introduced to a pond. Their numbers will climb rapidly as each cell divides into two new "daughter" cells. The two daughter cells then divide, leading to four granddaughter cells. The process continues, yielding eight cells, and then sixteen, and so on. Soon, the bacteria population is skyrocketing. But it's a process that can't continue forever. Limitations on food and space mean that at some point the bacteria population reaches an environmental limit. That limit is called the environments' carrying capacity. The population starts low, rises quickly, and then flattens out at the environment's carrying capacity.

A century later, Volterra (and others) took theoretical biology further by creating what is now known as the classic predator-prey model.[8] It begins with two equations. One tracks the prey population, which could be something like the number of bunnies in a forest. The second follows the predator population, which we could imagine as the number of wolves in the same forest. The important point for modelers to capture is that the two populations are tied together. The wolves eat the bunnies, and that changes the bunny population. But eating bunnies lets the wolves reproduce, adding to their population. So, the bunny population affects the wolf population, too. In these linked equations, there's a part (a "term") that describes how the bunnies get eaten by wolves, and another that describes how the wolves have more babies by eating bunnies.

In the language of math, the predator (wolf) and the prey (bunny) populations are coupled. They depend on each other. The two equations must be also solved together, which makes the problem tricky from a technical point of view. Volterra worked out this solution, and it showed him that the wolves and bunnies can end up cycling back and forth from high to low populations and back again. What was truly surprising, though, was the timing.

In an environment where both the bunny and wolf populations start low, the model predicted that only the prey begin increasing rapidly. The bunnies start reproducing first, and their numbers climb. The wolf population only begins increasing after enough bunnies are around to make them easy to find and catch.

Eventually, the bunny population peaks as the rapidly growing number of wolves starts having its impact. After that, the bunny numbers drop and they start to grow scarce. The wolf population, however, takes some time to feel the change. Only later do their numbers peak and then start dropping. Eventually, the wolf population gets low enough for the bunnies to recover, and the cycle begins anew.

What D'Ancona saw during the war was that the sharks (the predators) were still on the upswing, while the mackerel (the prey) were already past their peak and in decline. Volterra's model predicted the lag between population peaks, so it explained why the shark numbers would be seen increasing while the mackerel were falling. In this way, Volterra's theory—meaning his mathematical model—let D'Ancona get to the root of his apparent paradox.[9] The theory revealed the essential biology of predator-prey interactions.

What emerged from the work of Volterra and other pioneers was a true form of theoretical biology. In this setting, *theory* doesn't mean a hypothesis, like a detective's notion of who committed a murder. Rather, in science, theory means a large body of knowledge resting on mathematical principles that have been thoroughly verified through experience. The theory of population biology (also called population ecology) that Volterra and others founded was powerful enough that it could be applied to an ever-growing range of problems. Today, population biologists, ecologists, and their compatriots use mathematical models to study everything from the spread of disease to the propagation of invasive species.[10] Their approach would, eventually, even find its way to the study of human civilizations.

EASTER ISLAND, EASTER EARTH

Easter Island is a long way from anywhere. Located more than two thousand miles west of Chile and four thousand miles southeast of Hawaii, it's an isolated outpost of land surrounded by seemingly boundless expanses of ocean. The skilled Polynesian sailors who colonized the Pacific thousands of years ago didn't reach Easter Island in their long canoes until sometime around 400 CE. When they did, they found an island rich in fertile soil, as well as plant and animal life. It was a promising beginning to a story that would end in ruin.

When Dutch explorers discovered Easter Island on Easter Sunday in 1722, they found a "barren place with a few thousand people living in abject poverty and fighting over meager resources."[11] The island was devoid of trees, and the ground was covered only with unproductive scrub. But the huge stone monuments dotting the island and fashioned in the shape of human sentinels spoke of a very different past. Many of the "stone heads" were thirty feet high and weighed more than fifty tons. The silent faces of the monuments reflected a time when Easter Island had hosted a vibrant civilization with a population that may have peaked at more than ten thousand people.[12] Whatever culture existed before the Dutch arrived, it was technologically advanced enough to carve the monuments from rock located at the volcanic core of the island and transport them across miles of rugged terrain.

The mystery of what happened to Easter Island's civilization has haunted generations of writers and scientists. Erich von Däniken, in his 1973 bestseller *Chariots of the Gods?* went as far as to suggest an alien civilization was the only explanation.[13] How, he asked, could the islanders have moved the massive stone monuments when there were no trees around to use as rollers? But ancient aliens were not required. The answer to Easter Island's mystery turned out to be far simpler, and far more depressing.

There are no trees on Easter Island because the Easter Islanders cut them all down. They deforested their island in the building and transportation of those giant stone heads. In the process of deforesting the island, they also started a downward spiral that drove their civilization to collapse.

The iconic statues on Easter Island are evidence of a thriving civilization that collapsed before the Dutch landed there in 1722.

While there remains debate about the exact trigger for Easter Island's fall, environmental degradation driven by the inhabitants' own activity played an essential role. Easter Island serves as an object lesson for the interaction between an isolated, habitable environment and a civilization using that environment's resources: they did it to themselves. The parallel to our current situation on Earth seems clear.

In his 2007 bestseller, *Collapse*, anthropologist Jared Diamond unpacked that parallel.[14] His work explored the trajectories of a number of human civilizations that disappeared at the height of their vibrancy and power. Diamond's examples included the Anasazi of the American southwest, the Maya, and the Norse colony on Greenland. In each case, the civilization overshot the carrying capacity of its environment. Their populations grew as the society became ever more ingenious at extracting resources from its surroundings. Eventually, the limits to growth were hit. A short time after running into those limits, each civilization fell apart. Easter Island was the poster child for Diamond's story.

By the time Diamond brought historical examples of environmental collapse to the public's attention, scientists had already begun the mathematical modeling of Easter Island's fall. Using the same kinds of biological population models as those pioneered by Volterra and others, these researchers developed equations to explore the islanders' trajectory from vibrancy to collapse.

It began in 1995 with a paper by environmental economists James A. Brander and M. Scott Taylor.[15] Brander and Taylor set out two equations. The first described the change in the human population over time. The second described the change in the availability of the island's resources over time. Just as in Volterra's predator-prey models, the two equations were coupled. As the islanders used the island's resources for food and technology, their numbers grew. The resources, like trees, were renewable, and the equations could describe them growing back at natural rates, even as they were harvested by the islanders. But when Brander and Taylor solved their equations for the coupled trajectories of both the human population and the island's resources, their model tracked the islanders' fate with a grim certainty.

As the population grew, the resources could not keep up. Over-harvesting pulled resources down, and eventually, the island's inhabitants went with them. Peaking sometime around 1200 CE, the human population of Easter Island then experienced a gradual die-off, ending with just a few thousand inhabitants left by the time the Dutch arrived. The mathematical model got the general trend in the history right.

Other researchers soon followed up on Brander and Taylor's work. They changed the assumptions in the model by adding new terms to the equations or changing the form of the terms to reflect different kinds of interactions. A 2005 study by Bill Basener and David S. Ross[16] looked at the problem slightly differently. They assumed that the island had a carrying capacity for humans, as well as for the island's resources (like trees or animals). In their models, they then made the human carrying capacity explicitly dependent on the resources. As the resource levels declined, the ability of the island to host a human population would drop as well. When Basener and Ross solved these new equations for Easter Island's history, they saw something different from the gradual die-off Brander and Taylor found. The population climbed to its peak and then dropped like a stone—a true collapse.

Theory building regarding the history of Easter Island continues, with new studies appearing each year. There are many open issues that researchers must struggle with, since some of the data about the island before the Dutch arrival remains open to interpretation. But the basic path of the islanders' fate seems well captured in the models.

That success shows us the way forward for thinking about our own planetary fate in its proper cosmic context. What is true for an isolated island, its ecosystems, and its inhabitants should also be true for planets in the isolation of space.

A THEORETICAL ARCHAEOLOGY OF EXO-CIVILIZATIONS

In 1959, Carl Sagan took the greenhouse effect, a theory developed sixty years earlier for the Earth, and applied it to the distant planet Venus. In 1983, James Pollack and his collaborators took detailed models of dust storms on the distant planet Mars and applied them to Earth's own climate after a nuclear war. In the midst of the current exoplanet revolution, astronomers are taking knowledge gained from studying Venus, Mars, and Earth and applying it to the habitability of distant worlds orbiting distant suns.

For the last five decades, our knowledge of planets as generic cosmic phenomena has exploded. Data from these different worlds has been cross-pollinated with our understanding of Earth, helping us to understand other worlds, both in their own right and in relation to our own. This cross-pollination is so robust that scientists are now creating detailed models of possible biospheres on exoplanets. They want to be ready with predictions when soon-to-be-completed telescopes give them next-generation views of exoplanet atmospheres.

But if we are already creating theoretical models of biosphere-harboring exoplanets, what keeps us from carrying out the same process for worlds harboring civilizations? If we ask the right kind of questions, nothing stands in our way; we can get started now. By uniting our understanding of planets with population ecology—in the spirit of Volterra and those who followed—we can take a first stab at exploring the coupled trajectories of civilizations and their planets as generic cosmic phenomena.

It's a project that might be called a theoretical archaeology of exo-civilizations.[17] Anything we do concerning exo-civilizations will have to be theoretical. This is true not only because we don't have data, but also because our method will start from basic ideas about life and environments, as Volterra did in developing his predator-prey model. We want to let physics, chemistry, and population ecology guide us in unpacking the possible histories of exo-civilizations. Our goal with this theoretical archaeology of exo-civilizations is to see what could have happened to them, so that we can get a better handle on what might happen to *us*.

Given both the audacity and possible absurdity of anything calling itself a theoretical archaeology of exo-civilizations, let's boil the idea down to its core elements.

Step 1: Other Civilizations, Other Histories. As the pessimism line indicates, unless the universe has a really strong evolutionary bias against creating civilizations, we are not the first. If we are will-ing to take the existence of those other civilizations seriously, then we will recognize that each will have its own history in terms of interactions with its host planet.

Step 2: It's All about the Averages. We're really interested in things like Drake's final factor: How long, on average, does a tech-nological civilization last? That means the results of a single theo-retical model don't really tell us much. What we need are statistics compiled by modeling a large number of exo-civilizations. Thanks to the pessimism line, we know what that means.

Scientists usually like to have more than a thousand data points for whatever they're studying (this is true even in political polling). With that much data, quantities like averages make sense. So long as nature's choice for the biotechnical probability is one thousand times greater than the pessimism line, a thousand exo-civilizations will have already lived out their histories across cosmic space and time. Given the already tiny value of the pessimism line, it's not much of a leap to imagine that a thousand civilizations have already run their course. This would require a biotechnical probability of just one in ten thousand trillion (10^{-19}), which is still much smaller than most historical pessimists have feared.

Step 3: There Is No Free Lunch. Now we enter the territory where our astrobiological view of planetary science and climate studies comes into play. In the public debate about sustainability, the focus is often on switching our civilization's energy source from fossil fuels to something with less of a planetary impact. There is nothing wrong with such a goal, but the message often gets mangled in public debate from "less impact" into "no impact."

If we take the astrobiological view and start thinking like a planet, we see there's no such thing as "no impact." Civilizations are built by harvesting energy and using that energy to do work. The work can be anything from building buildings to transporting materials to harvesting more energy.

Without technology, each human being gets one human being's worth of energy each day. But with technology, we vastly expand the energy at our disposal. The average American uses the equivalent of about fifty servants just to power their home.[18] If we add in the energy needed for driving, flying, and other activities, the number of virtual servants gets much, much higher. Since this is just a matter of physics, what's true for us in terms of energy, power, and work must be true for any civilization-building species. The whole process of building a technological civilization is really an exercise in harvesting energy from the surroundings—in other words, from the planet.

So you can't build the kind globe-spanning, energy-intensive civilization we're interested in without having *some* impact on your planet. In fact, the laws of physics *demand* that you have an impact. Specifically, the Second Law of Thermodynamics is the culprit.

The Second Law tells us that energy can't be perfectly converted into useful work. There is always some waste. So any civilization-building species on any planet, using any form of energy, must produce waste. As that waste builds up, it turns into feedbacks on the planetary systems. From this perspective, the CO_2 produced by our burning fossil fuels can be seen as a kind of waste product of our civilization building. So, while the waste can take many forms, all of it will affect the planet. The states of the atmosphere, oceans, ice, and land will all change as the waste accumulates. That's the real scientific story of climate change and the Anthropocene.

Now, you might counter with the argument that civilizations more advanced than ours will find ways around the Second Law. Most physicists will tell you, "Good luck with that." The Second Law is baked into the structure of the universe, and being able to skirt it entirely is very unlikely.

But what capacities a highly advanced civilization might possess is an extremely important question for our theoretical archaeology project. It's so important, in fact, that our archaeology of exo-civilizations is designed explicitly to avoid speculating about it. And that leads us to the next step.

Step 4: Planets Come with a Limited Number of Energy Sources. In building our archaeology of exo-civilizations, we are going to focus explicitly on *young* technological civilizations. That means civilizations at our stage of development. This focus makes sense for two reasons. First, the whole point of this enterprise is to see what we can learn by treating our predicament as a general and generic phenomenon. The challenge humanity faces in the Anthropocene would not be so compelling and existential if we already had warp drives and other super-technology. Understanding our immediate fate is one good reason to keep our thinking focused on young civilizations. But the emphasis on youth is also essential for creating a project with strong scientific constraints.

One of the greatest impediments to thinking about exo-civilizations (or our own deeper future, for that matter) is technological progress. How can we anticipate what kind of technology a civilization that's a million years older might have at its disposal? Societies that mature might have found entirely new forms of energy that come from thin air. How can our theoretical modeling of exo-civilizations account for unknown sources of energy we haven't discovered?

The answer is, it can't. But luckily, it doesn't have to.

The development of technology is like climbing a ladder. You can't make a steel blade until you know how to make an iron blade. The Babylonians simply didn't have the capacity to build the metal-alloy components of a modern wind turbine. Each civilization must climb up the ladder of technological sophistication as it discovers the physical and chemical principles of the world around it.

For our project, that means a young civilization will have a limited number of energy sources available. Crucially, we know what those forms are. The laws of physics, chemistry, and planetary evolution tell us what resources might be at the ready for an intelligent species building its way up the technological ladder. Here is a pretty complete list of the energy resources a planet might offer:

- **Combustion.** This means burning stuff. It could be fossil fuels that are burned if the planet went through the right kind of geologic epoch, or it could just be biomaterials, like wood on our world.
- **Hydro/Wind/Tides.** If the planet has fluids or gases flowing on its surface, then those movements can be tapped to generate energy.
- **Geothermal.** Heat from the planet's interior can also be captured and used to do the work of civilization building.

- **Solar.** Sunlight can be trapped in both low-tech (heat) and high-tech (electric current) ways.
- **Nuclear.** The energy locked up in atomic nuclei can be used as long as there are reserves of radioactive elements like uranium around. Nuclear energy is obviously higher on the technological ladder than other modes of energy harvesting, but given that our society has made use of it, it's fair to think that others might as well.

The specific conditions on each planet will ultimately determine the mix of energy modes available to a civilization evolving there. Geothermal may be more favorable on some worlds, while wind may be more easily tapped on others. The main point for now is that the list above hits almost all the choices. Other than imagining exotic planets with special magnetic fields or continuous lightning conditions, what's on the list above is all that exists. Adding new energy sources other than those we've listed requires inventing science fiction stories about discovering "new physics."

Step 5: Know the Impact. Since we can list the different energy sources available to a young civilization, we can also calculate the planetary impact of their use. If this sounds like science fiction, remember that way back in 1903, Svante Arrhenius carried out exactly this kind of calculation for the Earth and combustion (that is, the burning of fossil fuels). Arrhenius knew the composition of the Earth's atmosphere, and he could calculate the impact of using coal. This impact was the production of CO_2, and the change it produced was an enhanced greenhouse effect.[19]

So, for civilizations powered by combustion, we already know how to model their impacts on their planets. All that's needed is to account for the potential differences in their host planets' properties, which will include things like the composition of the atmosphere and orbital location in the habitable zone.

What about the impact of other energy sources? For some cases, the calculations have already been started. A study by scientists at the Max Planck Institute in Germany looked at the global effects of wind power. Wind turbines work by pulling energy out of smooth, large-scale flows of air and turning it into electricity. But in the process, they leave choppy, turbulent airflows downstream. The German group found that extracting energy from wind power on a scale massive enough to power our current civilization would leave a global imprint akin to mild global warming. Even wind, the darling of renewable-energy harvesting modalities, has a planetary cost (though far lower than fossil fuels).[20]

Because we have a deep understanding of the physics and chemistry of each of the energy sources listed above, it doesn't take a quantum leap in science to calculate how their use will produce feedbacks on a planet other than our own. For each energy source a civilization might harvest, we have the information necessary to calculate the associated planetary cost. With that capacity, we reach the final step in the path to our theoretical archaeology of exo-civilizations.

Step 6: Turn the Crank. Given steps 1 through 5, we now have a recipe for calculating exo-civilization histories. We begin by creating a model for the interaction of a young civilization with its planetary environment. This model will come in the form of equations predicting how the civilization's population and its host planetary systems change with time. As in the predator-prey model, the equations will be coupled. There will be an equation describing the change in the planetary systems (such as atmosphere) and an equation describing the changes in the civilization-building population. Each equation will have terms that describe the feedback from planet to civilization and civilization to planet. It's worth mentioning that to do this job well, we'll need more than just two equations, because we'll probably need to track different resources and their use, along with their effect on the different planetary systems like oceans, ice, and so forth. But for now, we can stick with just "the planet" and "the civilization."

In general, the civilization will use its energy sources, and the waste from those energy sources will push on the state of the planetary systems. As the planetary systems shift based on the feedback, the civilization will either thrive or be stressed, as reflected in how their population changes. Because the coupling will be complicated, we won't know what to expect until we've solved the equations making up the model.

Doing this once doesn't tell us very much. What we are interested in is Drake's final factor: the average lifetime of civilizations. In order to calculate an average, we will have to run our models many times for many different kinds of planets. In a sense, by running the experiment in civilization building over and over, we will create our own mini-version of the universe. Some of the model runs will begin with planets that are close to the inner edge of their stars' habitable zones, where they'll be particularly susceptible to enhanced greenhouse warming. Some will be farther out. Some of our model runs will have planets with atmospheres that have less oxygen than ours, while others will have more oxygen. Some will begin with civilizations using wind power, and others will begin with civilizations using geothermal. You get the picture.

In the end, we will have to "turn the crank" and run tens of thousands of models, each with different starting conditions. That might seem like a lot of work, but modern computers are fast.

PATHS TO PROGRESS, ROADS TO HELL

Carrying out a theoretical archaeology of exo-civilizations correctly will be demanding. It will require input from fields as diverse as atmospheric science, geology, energy science, and ecology. To create realistic models, we'll have to get the physics, chemistry, planetary science, and ecological interactions right in terms of what we build into the models. That is going to be a long and interesting project.

But even as we build our way toward that goal, we can take some initial steps now. These first explorations can give scientists the lay of the astrobiological exo-civilization landscape. In the fall of 2016, a team of us went on this kind of scouting mission. The result was simultaneously thrilling, hopeful, and possibly a little depressing.

Our team included Marina Alberti, an urban ecologist from the University of Washington. A native of Italy, her passion is how evolution is already responding to the Anthropocene. Marina studies urban environments and how new species are being created in the midst of our vast project of planetwide city building. Axel Kleidon was also part of the effort. Axel is also an innovative thinker who works at the Max Planck Institute for Biogeochemistry, developing new ways to look at the Earth as a single thermodynamic system, like a giant, planetwide steam engine. Finally, there was Jonathan Carroll-Nellenback. Jonathan was my graduate student years ago and now works with me as a senior computational scientist at the University of Rochester. His talent for theoretical work is pretty remarkable. Sometimes I'll bring Jonathan a problem in the morning, and by the next day he'll bring it back, fully solved and displayed in beautiful graphics.

Together, we formulated a model for the evolution of a civilization with its planet. The equations were pretty simple. We weren't trying to capture the details of Earth or of any other specific planet. Instead, our aim was to describe the interaction of civilizations and planets in the most general way possible, which would serve as a first step toward doing something more detailed and realistic.

In our approach, the population and the environment were linked via an energy resource. The planet supplied the energy resource, and the civilization used the energy resource. Greater energy use meant a larger population on the one hand, and greater environmental change on the other. Greater environmental change lowered the planet's carrying capacity for the civilization, which should lead to lower populations.

Along with these features, we also included a specific mechanism to describe how the civilization might respond to changing conditions on its planet. For the sake of simplicity, we imagined that the planet had just two kinds of energy resources. One resource had a high planetary impact (as fossil fuels do), while the other had a low impact (as solar energy does). Here, high and low impact reflected the degree to which using the energy source forced the planetary environment to change.

Once the planetary environment was pushed past some pre-defined point, the civilization switched energy resources. You can think of this in terms of the planet's temperature. Once the planetary temperature rose to the specified value, the civilization stopped using the high-impact energy source and switched to the low-impact source.

Using this strategy in the models gave us a specific and simple way to boil down the civilization's sociology. We didn't want to try and model how they'd recognize and act on their Anthropocene. Instead, it came down to the planetary temperature that finally gets the civilization to do something. Since that was just an input, we could change it from one run to another and see how history played out for "smart" civilizations and "dumb" ones. Either the civilization acted early, when their planet's temperature had just started to rise, or they acted late. While we couldn't model the sociology of how they made that choice, we could model the choice's physical consequences. Would acting early save them? Would *anything* save them?

So, what did the model tell us?

Our exploration of the exo-civilization/planet system yielded three distinct trajectories. The first—and, alarmingly, most common—was what we called "the die-off." As the civilization used its energy resource, its numbers grew as expected (see page 196, graph A). But the use of the resource pushed the planetary environment away from its initial state. As the evolution of the coupled civilization/planet system continued, the population rose sharply beyond what the environment could sustain. The population, in other words, overshot what the planet could support. A big reduction in the civilization's population followed, until both the planet and the civilization had reached a steady state. After that point, neither the population nor the planet changed anymore. A sustainable planetary civilization was achieved, but at a considerable cost.

In many of the models, we saw as much as 70 percent of the population die before a steady state was reached. Imagine seven out of every ten people you know perishing because of global climate catastrophes. It's not clear how large of a die-off a complex technological society could handle without falling apart. During the period of the Black Death in the fourteenth century, Europe lost between 30 and 50 percent of its population, but managed to revive. Medieval Europe, of course, wasn't highly technological in the modern sense, nor as isolated as a planet in space would be.

The second trajectory class we found was one we called the

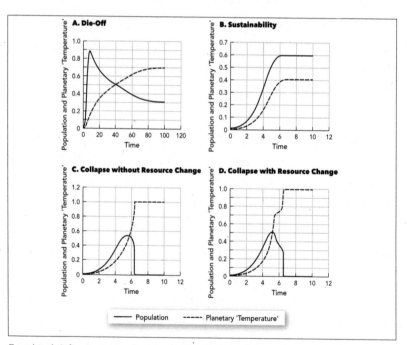

Four kinds of trajectories for exo-civilizations and their planets discovered from mathematical models.

"soft landing" (see page 196, graph B). The population grew and the planet changed, but the models showed a smooth transition to a steady state after an early switch to the low-impact energy resource. Eventually, the civilization came into equilibrium with its planet without a massive die-off.

The final class of trajectory was the most worrisome: full-blown collapse. As in the die-off models, the population initially grew swiftly. In this case, however, the speed of planetary change pulled the planet's carrying capacity down so fast that the population plummeted all the way to extinction.

One of the most remarkable aspects of this class was that the collapse was inevitable. One would think that switching from the high- to low-impact energy source would make things better. But for some trajectories, it didn't matter. If we used only the high-impact resource, the population reached a peak and then quickly dropped to zero (graph C). If we allowed the civilization to switch to the low-impact version of an energy resource, the collapse was only delayed. The population would start to fall, then appear to stabilize, and finally, suddenly, rush downward to extinction (graph D).

The collapses that occurred even when the civilization did the smart thing demonstrate an essential point about the modeling process: it can surprise you. Because the equations representing the model are complex, unexpected behavior can happen. These are consequences you wouldn't have thought of if you hadn't done the work of cranking out the solutions.

Only after you study the behavior seen in the models do you understand what happened. Remember that our simplified models were tracing the development of a civilization and its planet together. In the case of the delayed-collapse trajectories, we were finding scenarios that showed us that switching from a high- to low-impact energy resource won't matter if the change is made too late. Even though the civilization in our model recognized its entry into an Anthropocene-like transition and switched energy sources to make things better, the planet was already heading into new climate territory. Once the ball got rolling, the planet's own internal machinery took over. It wasn't coming back to the original climate state, and it took the civilization down with it as it ran away into a new state.

In these cases, the planetary environment's own dynamics were the culprit. Push a planet too hard, and it won't return to where it began. We know this can happen, even without a civilization present, because of what happened with Venus and its runaway greenhouse effect. Our models were showing, in generic terms, how a civilization could push a planet into a different kind of runaway through its own activity.

The work that Jonathan, Marina, Axel, and I did showed us some of the basic ways a civilization and its planet might change together. It was good that we saw that long-term sustainable versions of the planet-plus-civilization system were possible. But the warnings were there as well. The self-perpetuating feedbacks that drove some civilizations to collapse, even after they made the smart choices, were particularly sobering.

THE FINAL FACTOR

It's reasonable to ask what this archaeology of exo-civilizations really tells us about reality. Aren't these models just mathematical toys? Isn't it true that we have not a single instance of a civiliza-

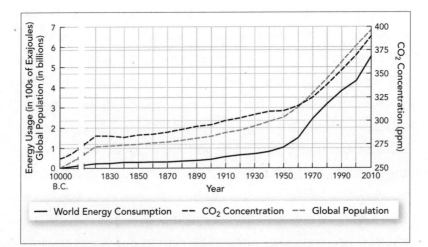

Models and Reality: The trajectory of the Anthropocene shown with real data for world energy consumption, CO_2 concentration, and global population for the last 10,000 years.

tion other than our own to make comparisons with? Answering these questions will help us see what can be gained by taking exo-civilizations seriously as subjects of scientific inquiry. It will also help us see what's at stake for us as we try to use this astrobiological perspective to understand our choices about our own project of civilization.

It is absolutely true that models and reality are two different things. A model is a simplification, like a skeleton without the muscle and skin. But looking at just a skeleton will tell you a lot about the animal. That is how we know about dinosaurs. More to the point, as we move forward, our models will be based on ever more sophisticated versions of what we already know about how planets work. They are, and will be, built on ever-stronger skeleton frames of physics and chemistry—in other words, the laws of planets. In that way, they are far more than mere imaginative toys.

The models allow us to go beyond fiction. By relying on the laws of planets, they capture key aspects of reality. That means they have their own logic. They have their own stories to tell us that we would not see without them. It's one thing to argue over what you think will happen when a civilization on a distant planet becomes technologically sophisticated. Your friend might have a different opinion, and that's an all-night argument waiting to happen. But it's something entirely different to spin up the math and let it *see* into the complexities that elude us. Instead of mere opinion, we can let the model show us how the universe might behave. The realistic constraints models place on their stories give those stories scientific value. It grounds them in the realm of the possible.

All the research we've explored in this chapter constitutes just a first step. It's an outline of what this kind of enterprise will look like as we devote more time and effort to the endeavor. The stories we've told here are just the first of many, and they will grow more precise as our understanding increases.

The next step will be to build far more realistic models and use them to explore a much wider range of realistic cases. After running these models for hundreds of thousands of different situations, we will have the simulated trajectories—the histories—of hundreds of thousands of inhabited worlds.

A planet that lives close to the inner edge of its habitable zone might be so highly sensitive to runaway greenhouse warming that its civilization barely has time to progress before it faces its own version of the Anthropocene and collapses. Another world, farther out from its star, may be less sensitive to planetary change but have a civilization that refuses to recognize the change until the die-off has already begun. A different species on a different world could manage to build its project of civilization using only lower-impact forms of energy and make a gentle soft landing to a sustainable state that lasts thousands of millennia.

What part of these stories matters to us? The answer to that question is simple: Drake's final factor. With trajectories for millions of simulated planets and civilizations in hand, we can calculate an average lifetime. How long, on average, does a civilization last?

Consider, for a moment, what that single number would tell us.

If the average lifetime of exo-civilizations is two hundred years, then we are in big trouble. If we find most model civilizations collapse after just a few centuries, the implication would be that civilizations like ours just don't work well on a planetary scale. A short average lifetime would mean that the universe doesn't do sustainable civilizations. The lesson would be that we humans are threading the eye of a needle with the Anthropocene and don't have much room for error in our choices. In that case, it may already be too late.

If the average lifetime of civilizations emerging from our models were tens of thousands of years, that would be good news. It would mean it's not too hard for any civilization to make it through the bottleneck of an Anthropocene.[21] There would be lots of different strategies for reducing our impact on the planetary systems that work. It would mean we have lots of wiggle room. We could make mistakes and still recover.

In this way, a single number from our archaeology of exo-civilizations—the average civilization lifetime—would have profound implications for our own future and our actions in the present. It would let us see what might be coming. And with that knowledge, our understanding of the choices we face would become deeper, richer, and be based on some wisdom.

Beyond the question of the average lifetime, we could also use the models to see exactly what choices are most likely to save us. Once we have a full suite of trajectories, we can ask what explicitly led some to civilizations to achieve planetary sustainability and others to collapse. Like a doctor looking for a cure by studying the most pathological cases of a disease, we can see what common factors drove the civilizations that died to their fate. The models will have a lot to teach us that we can't see now with the tunnel vision of just our planet and just our own uncertain future.

CHAPTER 6

THE AWAKENED WORLDS

THE PLANET WE NEED

If, over the course of billions of years of cosmic evolution, some species make it through their Anthropocene into long-term, sustainable versions of civilization, what do they end up with? What do their planets look like? How do these worlds function in terms of their coupled systems of air, water, rock, life, and the new addition of a planet-spanning, technology-intensive, energy-hungry society? These are the questions we care about most, because this is the target we must aim for.

There is a great deal of wishful thinking involved when the terms *planetary* and *sustainability* are parked next to each other. These are visions of "green utopias," with sleek, electric-powered trains gliding into elegant eco-cities of vertical farms and buildings mimicking natural forms. While it's easy to imagine what a single sustainable city might look like, imagining a sustainable planet is another thing entirely. Cities have always been the domains of

LIGHT OF THE STARS

human control. They are spaces our project of civilization carves out of nature. A planet, however, is a different beast.[1]

Planets are their own masters. That's what the astrobiological perspective shows us. The processes shaping worlds are powerful, complex, and subtle. Planets channel vast energies through ever more refined networks of cause and effect. These networks are embodied in winds that pick up fine dust grains and carry them across thousands of miles, or chemical compounds blown into the air by volcanoes, only to end up, millions of years later, embedded in rocks lying deep beneath oceans. Add life to the mix, and planets become almost infinitely more complex, as the planetary systems can now include a coevolving biosphere.

So, how does a healthy planet with a healthy, long-term project of civilization work? To answer this question, we must take our investigation to the final level. Crossing to the safety of a fully sustainable project of civilization on Earth requires not just thinking like a planet, but understanding the profound consequences of planets that have themselves learned to think through their civilizations. What, in other words, does it mean for a planet as a whole to wake up?

THE RUSSIAN MEETING

Ten years after their famous encounter at Green Bank, Frank Drake, Carl Sagan, and two other members of the original meeting found themselves together again. This time, the setting wasn't the forests of West Virginia, but a mountain in Armenia. Drake and his compatriots, along with a squad of Russian scientists, had come to the

Byurakan Observatory for the first true intraplanetary (or inter-national) meeting on interplanetary civilizations.[2] While Green Bank had been an intimate affair, with just nine members, the 1971 Byurakan Observatory meeting had more than forty participants, including luminaries from both the Soviet and American scientific establishments. There were Nobel laureates like Francis Crick (co-discoverer of DNA) and Charles Townes (inventor of the laser). Other notables included artificial-intelligence pioneer Marvin Minsky and Canadian neurophysiologist David Hubel, who would go on to win a Nobel Prize for his brain studies.[3]

Carl Sagan played a central role in organizing the Byurakan meeting. At the height of the Cold War, Sagan understood the symbolic value of an international conference devoted to our place among other, hopefully more mature, civilizations. Getting the

Carl Sagan (right) with Armenian astronomer Hrant Tovmassian (center) and other participants at the first international SETI meeting at the Byurakan Observatory in Armenia in 1971.

meeting placed in the Soviet Union, the United States' bitter enemy, had been no small task, however. To pull it off, Sagan needed a partner on the Russian side who was just as charismatic and passionate about life in the universe as he was. He found that counterpart in Nikolai Semenovich Kardashev.

Just a year and a half older than Sagan, Kardashev was a radio astronomer who had already made important contributions to the study of galaxies and the interstellar medium.[4] He had been the force behind the first Soviet search for exo-civilizations, carried out just a few years after Project Ozma. He'd also led the Soviets' first internal SETI conference in 1964, a few years after the Green Bank meeting. The report he had written on that meeting established his international reputation as a leading thinker about life and the evolution of other worlds.

In that paper, Kardashev laid out a scheme for the technological progress of exo-civilizations. His ideas played an important role in the Byurakan meeting, where freewheeling discussions of the long-term fate of civilizations went on until the small hours of the night. But the impact of what came to be called the Kardashev scale would last beyond the Byurakan meeting and would, in its way, prove to be as enduring as the Drake equation.[5]

THE KARDASHEV SCALE

When Nikolai Kardashev proposed his scale for measuring the progress of civilizations, he was primarily interested in finding them. Kardashev's question can be rephrased in a straightforward way: What are the milestones that mark a civilization's advancement up the ladder of technological sophistication? The basic idea that civilizations evolve through distinct, quantifiable stages as they progress offered Kardashev a lever to lift the discussion about exo-civilizations above pure speculation by providing a means to quantify their advancement. While his main interest was finding radio signals from exo-civilizations, his scale gave us a way to think about their evolution. But within the Kardashev scale lay an essential and mistaken bias concerning the relationship between civilizations and their host planets. Correcting that bias is an essential step in finding a wise astrobiology-based path through the Anthropocene.

Kardashev based his classification scheme on the energy a civilization had at its disposal. The scale had three levels.

- **Type 1:** These civilizations can harvest the entire energy resources of their home planet. In practice, this means capturing all the light energy that falls on the world from its host star, since stellar energy will likely be the largest source available on a habitable-zone planet. The Earth receives the equivalent of thousands of atomic bombs' worth of energy from the Sun every second.[6] A Type 1 species would have all this power at its disposal for civilization building.
- **Type 2:** These civilizations can harvest the entire energy resources of their home stars. The total output of the Sun every second is a billion times larger than the sunlight that falls just on Earth. The physicist Freeman Dyson anticipated some of Kardashev's thinking in a paper written in 1960, in which he imagined an advanced civilization constructing a vast sphere around its star.[7] This solar system–sized machine would capture stellar light energy, perhaps via an inner surface covered

in solar cells. Such "Dyson spheres" became the archetype for scientists imagining how a Kardashev Type 2 civilization would go about its energy-harvesting business.

- **Type 3:** These civilizations harvest the entire energy resources of their home galaxy. A typical galaxy contains a few hundred billion stars. Perhaps Type 3 civilizations envelop all their galaxy's stars within Dyson spheres, or perhaps they have even more exotic technologies at their disposal.

The Kardashev scale represents the scientific imagination working at the grandest, and therefore most mythic, scale. A single Dyson sphere would be a machine of staggering capacity and size. The inner surface of a Dyson sphere built around the sun, with a radius the size of Earth's orbit, would cover more than ten thousand trillion square miles (the equivalent of almost a billion Earths). Building a machine of this size would require grinding up whole planets for construction materials. We won't be building one of these anytime soon. Dyson spheres are truly the stuff of science fiction.

But by focusing on stellar-energy capture as the yardstick for a civilization's evolution, Kardashev's science fiction–sounding evolutionary scheme could be set firmly in the real world of real physics. That is what gave the Kardashev scale its reach, and why it has endured. For example, a number of researchers, such as Jason Wright at Penn State University, have conducted astronomical searches for the radiation signatures of Type 2 civilizations via their Dyson spheres.[8] Thus, as astronomer Milan M. Cirkovic wrote in 2015, "Kardashev's scale remains the most popular and cited tool for thinking about advanced extraterrestrial civilization."[9]

A large part of the appeal of the Kardashev scale lies in its combination of science and mythic-scale optimism via a technologically coherent road map for the progress of civilizations. Its implications are undeniably hopeful. If we continue to advance as a technology-building species, we should naturally pass through each Kardashev type on our way to a future of unimaginable power and reach. A civilization that could build a Dyson sphere would be the equivalent of technological demigods to us. In physics, power is defined as energy used per time. Since the Kardashev scale is explicitly based on energy use, the links between a civilization's physical power and its metaphorical power—between the science and the mythic—are baked into the scale's application. Make it far enough, the scale tells us, and you will become as the gods.

More than one author has tried to calculate where on the Kardashev scale human civilization falls today. In 1976, Carl Sagan suggested a way of calculating "fractional" values of Kardashev status based on world energy production.[10] In Sagan's calculation, we end up at about Type 0.7. Freeman Dyson went further, suggesting that human civilization will reach full Type 1 status in approximately two hundred years (with Type 2 requiring another hundred thousand to one million years).[11]

That sounds pretty good. In just a couple of centuries, we are going to become a true Type 1 cosmic civilization. The problem, of course, is that we may never get there. Our project of civilization has a bottleneck to navigate right now, and our progress through it is anything but assured.

The Kardashev scale originated from a particular historical moment in thinking about exo-civilizations. Like Sagan and Drake,

Kardashev was raised on a techno-utopian vision of the future. Technology was imagined in terms of sleek, gleaming machines that were destined to be humanity's salvation. We could expect that technology's growth and power would be unconstrained. That was why the Kardashev scale focused solely on energy. Civilizations were expected to rise up the ladder of energy harvesting to ever-greater heights until the entire galaxy would become a resource to be mined. And at each stage (each Kardashev type), the feedback from all this energy use on the physical systems from which the energy was drawn could be ignored. Planets, stars, and galaxies would all simply be brought to heel.

While it is possible that stars and galaxies might not care what you do with their energy, planets are another story. That is the painful lesson of the Anthropocene.

The engineering of entire solar systems or galaxies is so far into the realm of speculation that it's impossible to know what challenges it will require. But for planets—the focus of Type 1 civilizations—we already understand enough to see how the Kardashev scale represents a kind of planetary brutalism. It inherits a vision of advanced civilizations living in perfect, world-girdling cities where nature is fully controlled. Science fiction is full of this kind of thing. There is Trantor, the home world of the galactic empire in Isaac Asimov's classic Foundation trilogy. Trantor's surface lies hundreds of miles below the many shells of machinery that make up its single planet-scale city.[12] A more recent example is Coruscant, the home world of the galactic republic in *Star Wars*, with its continuous stream of "air cars" traveling amid the city's towering spires. These are visions of planets conquered by the mighty energy-wielding capacities of their civilizations.

But in the years since Kardashev proposed his classification system, we have learned the hard way that planetary biospheres are not so easily ignored. From the work of Lovelock, Margulis, and others, a new scientific understanding of planets and life emerged. Even when they lack life, we now know planets are complex systems. And if a vibrant biosphere is present, it becomes part of that complex whole. The living and non-living parts of the system coevolve across time. In this way, the coupled systems that make up a planet have their own internal dynamics—their own logic. That logic must be fully embraced when mapping out the trajectories of civilizations, as Kardashev hoped to do.

Once again, we are forced to stop seeing civilizations like our own as standing apart from the world that gave them birth. All civilizations, including those that might occur on other worlds, are expressions of their planet's evolutionary history. From this perspective, our project of civilization is just one consequence of the Earth's history, not its future master. Every civilization must be seen as a new form of biospheric activity arising within a planet's history of transformation and evolutionary innovation.

So it is not simply energy consumption (the focus of the Kardashev scale) that must be considered. Instead, we must learn to think in terms of energy transformations. We need to look at those physical laws that constrain energy as it flows through a planetary system. This means we must take a fully thermodynamic perspective as we follow the energy of sunlight being turned into the energy of rising air columns, which turns into the energy of falling rain, and so on, all the way to the energy of living cells.

Recognizing the limits on energy transformation is the fundamental lesson of the Anthropocene. You can't just bring a planet to heel, meaning you can't use energy to build a civilization without expecting feedback. Instead, we must begin with a richer understanding of biospheres and civilizations as part of the coupled planetary systems. That means a new kind of map for how civilizations rise to the Type 1 stage and, possibly, survive long enough to become something more. The development of long-term, sustainable versions of an energy-intensive civilization must be seen on a continuum of interactions between life and its host planet.

Sustainable civilizations don't "rise above" the biosphere, but must, in some way, enter into a long, cooperative relationship with their coupled planetary systems. But what does that look like?

EARTH AS A HYBRID PLANET

Planets are nature's way of turning starlight into something interesting. The evolution of a planet across billions of years depends on which processes it can harness to absorb starlight and, by doing work, transform that energy into something else. From rainstorms to forests to civilizations, the story of planetary evolution across cosmic time is the story of these energy transformations.

Energy flows are the domain of *thermodynamics*. The engine in your car is a thermodynamic system. It's a "heat engine." Gasoline gets ignited in the cylinders, converting chemical molecular-bond energy into hot gas, or heat energy. The hot, expanding gas pushes on the pistons, converting heat into motion via kinetic energy. The motion of the pistons gets transferred though the gears into the motion of the wheels.

So, it's not just the energy in the gas tank that matters. It's the transformation of that energy from chemical form into kinetic form that you need to pay attention to. Some of that original chemical energy gets dissipated (meaning it's lost and can't help in doing the work of moving the car) through the heating of the engine block or the friction of the tires on the road.

The science of thermodynamics tells us about the limits of those transformations. It tells us that not all the initial energy (contained in the fuel in the gas tank, for example) can be used to do useful work. Some of it, by necessity, must turn into "waste." Nature has built these limitations into the universe through the laws of thermodynamics.[13] That is why thermodynamics is the right way to think about planets and civilizations and their combined fate.

For a planet with no atmosphere, like Mercury, the available energy transformations are pretty limited. Sunlight hits Mercury's surface. The surface warms up and emits heat radiation back into space. Once the planet's surface reaches its equilibrium temperature, there is not a whole lot more to the story, which is why Mercury has looked pretty much the same from one day to the next for the last three billion years or so.[14]

Add an atmosphere to the planet, however, and the story gets a lot more interesting. When sunlight warms the surface of a planet with an atmosphere, the air near the ground is also warmed. Then the air rises, creating large-scale "convective" circulation. Atmospheric gas rises, and then cools and falls back toward the surface to start the circulation over again. Atmospheric convection is a kind of planetary heat engine, converting sunlight into motion.

If the atmosphere also includes molecules like water, CO_2, and other "volatiles," then evaporation and condensation can occur in the circulation.[15] Water, for example, will evaporate near the planet's surface, turning into a gas that rises with the rest of the air. When the air cools at higher altitudes, the water condenses back into a liquid (in the form of droplets). This is how interesting things like rain or snow can occur—things that could not happen on an airless world.

Just these ingredients—an atmosphere with stuff that can evaporate and condense—are enough to give a planet climate and weather. It's why even a relatively "dead" world like Mars can still look different from one day to the next, as dust storms, fog, or frost roll in and out.

The presence of liquids flowing across the surface, in the form of rain runoff and rivers, adds a new layer of "interesting," as the strong weathering of rocks can begin. Elements once locked up in minerals get exchanged with the air and the surface liquids, launching "cycles" of these materials between the planetary systems.[16] The branching pathways of these cycles and their feedbacks bestow a new richness to the planet, allowing it to evolve in even more complex ways.

The point here is that all of these processes are fundamentally transformations of energy. The presence of an atmosphere turns solar energy into motion energy as air rises and falls. With water or CO_2 in the atmosphere, the energy of motion feeds into the energy associated with evaporation and condensation. Weathering and the breaking of chemical bonds in rocks is yet another form of energy transformation. So, even without life, a planet can take its sunlight and use it for ever more complex work, driving change, evolution, and innovation.

Thinking this way about evolution and energy led Marina Alberti, Axel Kleidon, and me to propose a new classification for planets.[17] While the Kardashev scale focused on the total energy falling onto a planet, we were interested in what happens to that energy once it gets within a planet. Here, "within" doesn't mean underneath the surface of a world, but within the coupled planetary systems. What happens when a sun's energy, in the form of incoming light, feeds through the linked networks of atmosphere, hydrosphere, and so on, including a biosphere?

Unlike Kardashev, our goal in making a new planetary classification scheme was not detection (though it proves useful for this). Instead, we wanted to use the laws of physics, chemistry, and biology on planetary scales to see where planetary evolution might lead. In particular, we wanted to use the planets we already understood to map out the properties of the ones we don't understand—the planets with sustainable civilizations.

Working together, the three of us saw that the universe's vast census of planets might be grouped into a spectrum of five main classes.

An airless world like Mercury is a Class 1 planet in our scheme. The transformations of sunlight are simple, and so the degree of work that is done and the complexity generated are limited. Class 1 planets are truly dead worlds.

A world with an atmosphere but no life, like Mars or Venus, is a Class 2 planet. The flow of gases and liquids driven by sunlight represent work being done within the planetary systems. That work can make things happen on a range of time scales, like the daily appearance of fog or the yearly appearance of dust storms.

Class 3 planets are those with what we called a "thin" biosphere.

These are worlds where life has gotten started. It's affecting the rest of the coupled systems, but does not yet dominate these systems. One way to quantify this is to look at what's called the net productivity of a planet, meaning how much energy its biosphere harvests. Donald Canfield has estimated the net productivity of the Earth's early Archean biosphere and found that it was a hundred times smaller than today.[18] So Earth during the Archean was a Class 3 world. If Mars had life during its wet Noachian period, four billion years ago, then it too might have been a Class 3 world.

Class 4 planets, on the other hand, have been hijacked by life. They have "thick" biospheres that are deep networks of animal, plants, and microbes, all feeding on each other and all feeding back onto the other planetary systems. The existence of our oxygen atmosphere, created in the Great Oxidation Event, tells us that we are living on a biosphere-dominated planet where life plays an outsized role in planetary evolution. So the Earth, before civilization appeared ten thousand years ago, was a Class 4 world.

Our scheme was based on the fact that we had real examples of the first four planetary classes. Through these known worlds, we could understand how solar energy feeds through the planetary systems and drives evolution. That knowledge gave us purchase to see something essential about our hypothesized fifth class: a world hosting a sustainable civilization.

Going from Class 1 to Class 4 worlds, we see an increase in the complexity of their energy flows and transformations. Class 1 worlds could do little in terms of turning solar energy into work and change. Class 4 worlds comprised rich networks of processes channeling solar energy into work and change. From the perspective of thermodynamics, we could see how planets in each successive class had "found" new ways to transform their incident starlight into evolution. On a world without life, this evolution can be rich, but the pathways are constrained purely by physics and chemistry. In a sense, its details are fairly predictable. Once life appears in Classes 3 and 4, biological evolution takes over. Life figures out entirely new ways to do work, yielding new processes that feed back on the rest of the planet.

The relationship between complexity, work, and energy flows gave us a key to understanding what our fifth class of planet might look like. A thick biosphere on a Class 4 world channels more energy into work than a thin biosphere on a Class 3 world, which itself channels more energy into work than on a Class 2 world. That means a planet with a sustainable civilization—a Class 5 world—might be even more adept at wringing work and change out of sunlight. On a Class 5 planet, the biosphere—which now includes a globe-spanning civilization—becomes even more productive than Class 3 and Class 4 worlds. The civilization not only harvests more energy, as Kardashev imagined, but also figures out how to put this energy to work in ways that do not push the planet into dangerous territory. The civilization, as part of the biosphere, adds what philosophers call "agency." The civilization makes choices with goals in mind. Thus, Class 5 planets have *agency-dominated* biospheres. The civilization is now deliberately working with the rest of the natural systems to increase the flourishing and productivity of both itself and the biosphere as a whole.

Perhaps the civilization converts its planet's deserts into productive ecosystems. Such "desert greening," if done correctly, could stabilize a changing climate. Or it might engineer plants that can both photosynthesize and produce electricity (there are researchers studying this now).[19] Or it might cover regions with solar cells in ways that also increase (or at least don't decrease) the total biospheric productivity and health of the planet. The possibilities are rich, and our study was meant only to suggest the right direction a Class 5 agency-dominated biosphere might take. There is much fruitful work to be done in turning the basic concept of Class 5 worlds into strategies for the future.

So, where does Earth fit into our classification scheme right now? As we enter the Anthropocene, we are clearly leaving the Class 4 state. Our activity and choices are strongly modifying the state of the biosphere and other planetary systems. But we are making these changes without a long-term plan, as planetary scientist David Grinspoon and others have pointed out.[20] We are evolving the planet toward something new, but we can't say if that novel state will include us in the long term. So, Earth at the beginning of the Anthropocene is no longer a Class 4 world but is not yet, and may never be, a Class 5 planet. As of now, it's a hybrid world. It's evolving toward something other than it was, and it's doing so in a way that's dangerous for our project of civilization.

The key point in developing these five classes of planet was the necessity of putting civilizations back into the context of the biosphere, rather than above it. From this perspective, sustainable civilizations are extensions of the long process of planetary evolution. Biospheres without civilizations are already agents of novelty. From oxygen-producing microbes to grasslands to megafauna (like wooly mammoths), they produce new things that then enter into the web of positive and negative feedbacks on the planetary system as a whole. The great lesson of Lovelock, Margulis, and their Gaia theory was that the biosphere could evolve feedbacks that kept the system stable. A sustainable agency-dominated biosphere should be no different.

After his pioneering work on the biosphere, Vladimir Vernadsky went on to consider the possibility of planets "waking up" via what he called a "noosphere." Coined from the Greek *noos*, for intelligence, a noosphere was a shell of thought surrounding the planet. It was the result of a biosphere evolving creatures that could think and develop technology. From geology to life to mind, the emergence of the noosphere was, for Vernadsky, a next stage in planetary evolution.[21]

Class 5 planets might be seen as worlds that have evolved a noosphere. The pervasive wireless mesh of connections that constitute today's internet has already been held up as an initial version of a noosphere for Earth. Thus, we might already make out the contours of what a sustainable world will look like. To truly come into a cooperative coevolution with a biosphere, a technological civilization must make technology—the fruit of its collective mind—serve as a web of awareness for the flourishing of both itself and the planet as a whole.

Beyond the Kardashev scale's focus on energy as the currency of planetary dominance, we now encounter an essential lesson the stars might teach us about our next moves. *Planets are engines of innovation.* But, from Class 1 to Class 4, those innovations are blind. They are the result of pure chance and pure mechanics—the laws of physics, chemistry, and biological evolution. They do not have an end in mind. There is no teleology.

Recall that one of the loudest criticisms of Gaia theory was that it could be interpreted to imply that life on Earth "wanted" to steer the planet in some direction. It was in response to these criticisms that Gaia morphed into the less controversial Earth systems theory. There, evolution was once again blind. But when a civilization emerges and triggers its own version of the Anthropocene, the age of blindness must come to an end.

In the deepest sense, Class 5 planets would represent the completion of Gaia. They would be worlds where the planet as a whole has an evolutionary direction, a goal. That is what an agency-dominated biosphere means. The civilization, working for its own continued existence, recognizes itself as an expression of the biosphere and chooses a direction.

So, we cannot bring the world to heel. Instead, we must bring it a plan. Our project of civilization must become a way for the planet to think, to decide, and to guide its own future. Thus, we must become the agent by which the Earth wakes up to itself.

THE WAY FORWARD

Ultimately, the problem we face is confronting a twenty-third-century dilemma armed only with a thirteenth-century mind. Our project of civilization has been successful on scales we could not have imagined when we began it ten millennia ago. But with that success has come consequences that will last for centuries.

Across the long history of our project, we didn't know our true place in the universe and could not, therefore, know our place within the planet's own evolution. But now, through science, we can see a new truth. The Earth is but one world among trillions, and we are not a one-time story. Now we can—and must—make this our story.

We must make it the human story, one that cuts across cultures, nations, and politics.[22]

We are, most certainly, not the first species that has dramatically changed the Earth's climate. It has happened before, and we can see how that story played out in the past. Earth is possibly, and even likely, not the first planet that has evolved a civilization. Using all we have learned about planets, we can see how that story, including climate change, might also have played out in the past.

But what the Anthropocene means for the planet, and what it means for us, are different things. If we continue to do nothing about our use of fossil fuels and the other drivers of the Anthropocene, it is more than conceivable that we'll push the planet into domains that prove difficult for our kind of complex global civilization. If our project of civilization collapses for a time, or even permanently, the Earth will happily move on without us. In that sense, our urgency in dealing with climate change and the Anthropocene has nothing to do with "saving the planet." Our entry into the Anthropocene shows that our project of civilization has now become its own kind of planetary power. It's a new story we have to tell about ourselves, and everything now depends on learning and acting upon it.

Across the pages of this book, we've assembled this new narrative through smaller stories of that story's own evolution. We have encountered heroic scientists who took us up to the mountain so that we might see farther. There were Frank Drake, Jill Tarter, and Nikolai Kardashev, who braved the scorn of their colleagues to take the existence of exo-civilization seriously as a topic for scientific inquiry. Through their efforts, we could begin to see life and the stars in a new light. There were explorers like Jack James and Steven Squyers, blasting robots across space to the other worlds in our solar system. Through their work and the studies of researchers like Robert Haberle, we learned the laws of climate and evolution for all planets. Army corpsmen and scientists like Willi Dansgaard braved Camp Century on Greenland's ice sheet to help us see more deeply into the transitions of Earth's climate. Then came people like Donald Canfield, who traveled the world to unpack the deep history of our planet and its life. Putting all this together were visionaries like Vladimir Vernadsky, James Lovelock, and Lynn Margulis, who lifted our sights to see how that life can partner with its planet to evolve into something greater, something more. Finding other planets was the job of scientists like Michel Mayor, Bill Borucki, Natalie Batalha, and others. Their work answered a millennia-old question and, in doing so, filled the night sky with a trillion trillion worlds and possibilities. And finally, appearing at almost every turn, there was Carl Sagan. More than almost anyone else, we owe the possibility of this new story to his genius.

Science has given us a new perspective, a new vision, and a new story that can help us find a way forward as we face the challenge of the Anthropocene. But this can only happen if we listen carefully and truly make this new story our own.

It is time to grow up.

The central argument of this book, and one that Carl Sagan already understood, is that humanity and its project of civilization represent a kind of "cosmic teenager." We are likely just one world among many that has grown a civilization to the point where it has gained power over itself and its planet. But, like a teenager, we lack the maturity to take full responsibility for our ourselves and our future.

Gaining the astrobiological perspective is the first, essential step in our maturation and our ability to face the Anthropocene.

It means recognizing that we and our project of civilization are nothing more than the fruit of Earth's ongoing evolutionary experiments. Any civilization on any planet will be nothing more than an expression of its home world's creativity. We are no different from those we would call "alien."

So our focus has to shift. It's time to leave the tired question, "Did we create climate change?" behind. In its place we must take up our bracing new astrobiological truth: "Of course we changed the climate." We built a planet-spanning civilization. What else would we expect to happen?

But we should also recognize that creating climate change wasn't done with malevolence. We are not a plague on the planet. Instead, we *are* the planet. We are, at least, what the planet is doing right now. But that is no guarantee that we'll still be what the planet is doing one thousand or ten thousand years from now.

As children of the Earth, we are also children of the stars. If nothing else, the Anthropocene can make that fact as real to us as the shriek of a howling storm, the oppressive heat of a desert landscape, or the cool silence of a deep forest. Through the light of the stars, through what they can teach us about other worlds and the possibilities of other civilizations, we can learn what path through adolescence we must take. And in that way, we *can* reach our maturity. We can reach our full promise and possibility. We can make the Anthropocene into a new era for both our civilization and the Earth. In the end, our story is not yet written. We stand at a crossroads under the light of the stars, ready to join them or ready to fail. The choice will be our own.

ACKNOWLEDGMENTS

This work would not have been possible without the support and, sometimes, direct intervention of some very smart and kind people. I must first thank Howard Yoon of Ross Yoon Agency for his working so closely with me to develop the early versions of the idea into a coherent form and his many years of help in so many, many ways. From the very beginning, my editor at W. W. Norton, Matt Weiland, saw how to make the idea and its incarnation into this book cleaner and more sharply defined. It was a great pleasure to work with him and I am deeply grateful that his skills were brought to bear on this project. Simply put, he is a great editor. I am also grateful to have had Remy Cawley on the W. W. Norton team. In editing, copyediting, and managing the image process, her precision and thoroughness were essential. I was also lucky to have two wonderful University of Rochester undergraduates working with me as assistants on the book. Molly Finn worked tirelessly on fact checking and accumulating proper references and reference forms. Elise Morgan endured a crazy autumn tracking down images and permissions.

Both showed remarkable skills for young scholars, and I was lucky to have found their help.

For a scientist, it is always a little frightening to write about topics that are not squarely in the domain of your research specialty. For me, this included not only sciences like atmospheric chemistry, but also the amazing history of the discoveries I wanted to explore in the book. I bear full responsibility for any mistakes made in the text. In trying to get the story right, though, I was helped by many scientists who were generous with their time. In particular, my collaborator Woody Sullivan at the University of Washington provided many fine insights into the manuscript. My gratitude to him runs very deep. Jason Wright of Penn State gave an early version of the book a thorough and deep reading. Not only did he make the book more accurate, but he also drove me to think more deeply about a number of topics on exo-civilizations. Jill Tarter not only gave me a number of great interviews, but also provided excellent feedback on the manuscript. I am equally grateful to Donald Canfield, both for interviews on atmospheric chemistry and his review of the chapter on Earth science. James Kasting of Penn State and Lee Murray of the University of Rochester provided excellent feedback on the climate and Earth science sections as well. Robert Haberle was generous with his time in explaining the history of Mars climate modeling and reviewing the chapter on solar system exploration.

I am also grateful to Soren Gregersen, whom I bothered a number of times to tell me his stories of being a Boy Scout and living out on the Greenland ice sheet with the US military at Camp Century. I am equally grateful to Natalie Batalha and Bill Borucki for giving me their time for interviews.

ACKNOWLEDGMENTS

There are also many people I have to thank just for their intellectual companionship. Robert Pincus and Paul Green are always at the top of the list on any topic for me. My PhD advisor and continuing collaborator Bruce Balick has always been a source of good advice and ideas. I also had many excellent conversations with my colleagues at the University of Rochester: Dan Watson, Eric Blackman, Alice Quillen, Eric Mamejek, Judy Pipher, and Bill Forrest. Also, I must thank my collaborators on the work described in this book: Woody Sullivan, Marina Alberti, Axel Kleidon, and Jonathan Carroll-Nellenback. Ongoing discussions with Evan Thompson were also fun and helpful. Writing on these topics for both NPR and the *New York Times* gave me a first chance to cast the ideas in non-scientific language. I am very grateful to Jamie Reyerson at the *Times*, as well as Meghan Sullivan and Justine Kenin at NPR.

Finally, I am particularly grateful for my NPR blog co-founder, collaborator, and friend Marcelo Gleiser, who provided the opportunity to spend time at the Institute for Cross Disciplinary Engagement at Dartmouth, where some of this book was written. Thank you, Marcelo.

Finally, I must thank my children, Sadie and Harrison, as well as my brother-in-law, Hendrik Helmer, for making me laugh . . . a lot. And always, always, always, I thank the stars for my wife, Alana Cahoon, without whom none of this would matter.

NOTES

INTRODUCTION: THE PROJECT AND THE PLANET

1. John D. Durand, "Historical Estimates of World Population: An Evaluation," *PSC Analytical and Technical Reports*, no. 10 (1974): table 2.
2. Department of Economic and Social Affairs, Population Division, *The World at Six Billion* (New York: United Nations Secretariat, 1999), http://www.un.org/esa/population/publications/sixbillion/six billion.htm.
3. Paul Mann, Lisa Gahagan, and Mark B. Gordon, "Tectonic Setting of the World's Giant Oil and Gas Fields," in *Giant Oil and Gas Fields of the Decade, 1990–1999*, ed. Michel T. Halbouty (Tulsa, OK: American Association of Petroleum Geologists, 2014).
4. Department of Economic Affairs, Population Division, *World Population Prospects: Key Findings and Advance Tables, 2015 Revision* (New York: United Nations, 2015), https://esa.un.org/unpd/wpp/Publications/Files/Key_Findings_WPP_2015.pdf.
5. International Air Transport Association, *2012 Annual Review*, June 2012.
6. Lynn Margulis, "Gaia Is a Tough Bitch," in *The Third Culture: Beyond the Scientific Revolution*, ed. John Brockman (New York: Simon and Schuster, 1995).
7. Kim Stanley Robinson, *Aurora* (New York: Orbit, 2015).

8. University of Zurich, "Great Oxidation Event: More Oxygen through Multicellularity," *ScienceDaily*, January 17, 2013, www.science daily.com/releases/2013/01/130117084856.htm.

9. European Space Agency, "Greenhouse Effect, Clouds and Winds," *Venus Express*, http://www.esa.int/Our_Activities/Space_Science/ Venus_Express/Greenhouse_effect_clouds_and_winds.

10. V.-P. Kostama, M. A. Kreslavsky, and J. W. Head, "Recent High-Latitude Icy Mantle in the Northern Plains of Mars: Characteristics and Ages of Emplacement," *Geophysical Research Letters* 33, no. 11 (2006), doi: 10.1029/2006GL025946, and NASA Jet Propulsion Laboratory, "Mars Ice Deposit Holds as Much Water as Lake Superior," news release, November 22, 2016, https://www.jpl.nasa.gov/news/ news.php?release=2016-299.

11. Joe Mason and Michael Buckley, "Cassini Finds Hydrocarbon Rains May Fill Titan Lakes," Cassini Imaging Central Laboratory for Operations, January 29, 2009, http://ciclops.org/view.php?id=5471&js=1. The liquids on Titan include components of gasoline.

12. Colin N. Waters et al. "The Anthropocene Is Functionally and Stratigraphically Distinct from the Holocene," *Science* 351, no. 6269 (January 8, 2016), http://science.sciencemag.org/content/351/6269/ aad2622.

13. Dale Jamieson, *Reason in a Dark Time* (New York: Oxford University Press, 2014).

14. NASA Exoplanet Science Institute, "Exoplanet and Candidate Statistics," *NASA Exoplanet Archive*, https://exoplanetarchive.ipac.caltech .edu/docs/counts_detail.html.

CHAPTER 1: THE ALIEN EQUATION

1. C. P. Snow, *The Physicists* (Boston: Little Brown, 1981).

2. Alan Lightman, *A Sense of the Mysterious: Science and the Human Spirit* (New York: Vintage, 2006).

3. Eric M. Jones, *Where Is Everybody?: An Account of Fermi's Question* (Los Alamos, NM: Los Alamos National Laboratory, 1985), https:// www.osti.gov/accomplishments/documents/fullText/ACC0055.pdf.

4. Jones, *Where Is Everybody?*: 3.

5. Enrico Fermi, "My Observations During the Explosion at Trinity on July 16, 1945," *Fermat's Library*, http://www.atomicarchive.com/Docs/Trinity/Fermi.shtml.

6. As astronomer Jason Wright puts it, "Astronomers stare at the sky professionally with some of the most sensitive equipment in the world. If UFOs were common, we would see them all the time. It strains credulity that armies of amateurs with cameras regularly see UFOs when the professionals with giant telescopes do not." Jason Wright, "Astronomers and UFOs," *AstroWright*, December 1, 2013, https://sites.psu.edu/astrowright/2013/12/01/astronomers-and-ufos/.

7. Michael Hart, "An Explanation for the Absence of Extraterrestrials on Earth," *Quarterly Journal of the Royal Astronomical Society* 16 (June 1975): 128. Also see Robert H. Gray, "The Fermi Paradox Is Neither Fermi's Nor a Paradox," *Astrobiology* 15, no. 3 (March 2015): 195–99.

8. Glen David Brin, "The 'Great Silence': The Controversy Concerning Extraterrestrial Intelligent Life," *Quarterly Journal of the Royal Astronomical Society* 24, no. 3 (1983): 283–309, and James Annis, "An Astrophysical Explanation for the 'Great Silence,'" *Journal of the British Interplanetary Society* 52 (1999): 19–22.

9. Robin Hansen, *The Great Filter—Are We Almost Past It?*, September 15, 1998, http://mason.gmu.edu/~rhanson/greatfilter.html.

10. Heike Langenberg, "Slow Gulf Stream During Ice Ages?," *Nature News*, December 9, 1999, http://www.nature.com/news/1999/991209/full/news991209-10.html.

11. This could happen in many ways, but the easiest to imagine is a significant population reduction—a "die-off"—that keeps the population's capacities below the level for a technological/industrial re-emergence. Note that dramatic climate change could result in a species that once had a technological civilization living for hundreds of thousands of years on a world where large-scale agriculture has become impossible. It is very difficult to predict what the evolutionary/sociological outcome of this scenario would be.

12. Matthew F. Dowd, "Fraction of Stars with Planetary Systems, f_p, pre-

1961," in *The Drake Equation*, ed. Douglas A. Vakoch and Matthew F. Dowd (New York: Cambridge University Press, 2015), 56.

13. Steven J. Dick, *Plurality of Worlds: The Extraterrestrial Life Debate from Democritus to Kant* (Cambridge: Cambridge University Press, 1984), 6.

14. Dick, *Plurality of Worlds*, 26–27.

15. Dick, *Plurality of Worlds*, 62.

16. There remains some dispute over what exactly Bruno was convicted for in the heresy charge. Evidence points to more arcane issues of doctrine, rather than astronomy. His views on Copernicanism and other worlds, however, contributed to his career of conflict with the Church. Dorothea Singer, *Giordano Bruno: His Life and Thought* (1950; repr., New York: Greenwood Press, 1968).

17. Bernard de Fontenelle, *Conversations on the Plurality of Worlds* (1686; repr., London: J. Cundee, 1803), 112.

18. Dowd, "Fraction of Stars," 67, and Steven J. Dick, *Life on Other Worlds: The 20th-Century Extraterrestrial Life Debate* (Cambridge: Cambridge University Press, 1998).

19. Douglas A. Vakoch, ed., *Astrobiology, History, and Society: Life Beyond Earth and the Impact of Discovery* (Berlin: Springer, 2013), 108.

20. Percival Lowell, "Observations at the Lowell Observatory," *Nature* 76 (1907): 446.

21. William Whewell, *Of the Plurality of Worlds* (1853; repr., Chicago: University of Chicago Press, 2001), 207.

22. Whewell, *Plurality of Worlds*, 204–5.

23. Alfred Russel Wallace, *Man's Place in the Universe: A Study of the Results of Scientific Research in Relation to the Unity or Plurality of Worlds* (London: Chapman and Hall, 1904).

24. Dowd, "Fraction of Stars," 67.

25. Florence Raulin Cerceau, "Number of Planets with an Environment Suitable for Life, n_e, Pre-1961," in *The Drake Equation*, eds. Douglas A. Vakoch and Matthew F. Dowd (New York: Cambridge University Press, 2015), 98.

26. Natural Resources Defense Council, "Global Nuclear Stockpiles,

1945–2006," *Bulletin of the Atomic Scientists* 62, no. 4 (July/August 2006): 64–66, http://media.hoover.org/sites/default/files/documents/GlobalNuclearStockpiles.pdf.

27. Stephanie Pappas, "Hydrogen Bomb vs. Atomic Bomb: What's the Difference?," *Live Science*, January 6, 2016, https://www.livescience.com/53280-hydrogen-bomb-vs-atomic-bomb.html.

28. Don P. Mitchell, "The R-7 Missile," http://mentallandscape.com/S_R7.htm.

29. Steve Garber, "*Sputnik* and the Dawn of the Space Age," National Aeronautics and Space Administration, last modified October 10, 2007, https://history.nasa.gov/sputnik/.

30. Frank Drake and Dava Sobel, *Is Anyone Out There?* (New York: Delacorte Press, 1992), 5.

31. Drake and Sobel, *Is Anyone Out There?*, 27.

32. Drake and Sobel, *Is Anyone Out There?*, 8–12.

33. Frank Drake, "A Reminiscence of Project Ozma," *Cosmic Search* 1, no. 1 (1979): 10.

34. F. Ghigo, "The Tatel Telescope," National Radio Astronomy Observatory, http://www.gb.nrao.edu/fgdocs/tatel/tatel.html.

35. Drake, "Reminiscence."

36. John R. Percy, "The Nearest Stars: A Guided Tour," Astronomical Society of the Pacific, 1986, https://astrosociety.org/edu/publications/tnl/05/stars2.html.

37. Drake, "Reminiscence."

38. "Early SETI: Project Ozma, Arecibo Message," SETI Institute, http://www.seti.org/seti-institute/project/details/early-seti-project-ozma-arecibo-message.

39. Drake, "Reminiscence."

40. "Early SETI: Project Ozma."

41. Giuseppe Cocconi and Philip Morrison, "Searching for Interstellar Communications," *Nature* 184, no. 4690 (September 19, 1959): 844–46.

42. Drake and Sobel, *Is Anyone Out There?*, 32.

43. Drake and Sobel, *Is Anyone Out There?*, 45–64.

44. Drake and Sobel, *Is Anyone Out There?*, 47.

45. Drake and Sobel, *Is Anyone Out There?*, 54.

46. Drake and Sobel, *Is Anyone Out There?*, 49.

47. Drake and Sobel, *Is Anyone Out There?*, 51.

48. Maggie Masetti, "How Many Stars in the Milky Way?," *Blueshift*, July 22, 2015, https://asd.gsfc.nasa.gov/blueshift/index.php/2015/07/22/how-many-stars-in-the-milky-way/.

49. Fred Hoyle, *The Black Cloud* (London: Heinemann, 1957).

50. Drake chose to focus just on our home galaxy, the Milky Way, because the distances to other galaxies are so large. Any source-emitted electromagnetic radiation becomes more difficult to detect the farther away it is.

51. Su-Shu Huang, "The Problem of Life in the Universe and the Mode of Star Formation," *Publications of the Astronomical Society of the Pacific* 71, no. 422 (October 1959): 421–24.

52. Stanley L. Miller, "A Production of Amino Acids under Possible Primitive Earth Conditions," *Science* 117, no. 3046 (May 15, 1953): 528–29.

53. Drake and Sobel, *Is Anyone Out There?*, 61.

54. Of course, one can also ask whether a civilization that was far more advanced than ours would still use radio at all. But like the previous issue of life requiring planets, one has to begin somewhere, and its better to underestimate the likelihood of each term than go overboard.

55. Drake and Sobel, *Is Anyone Out There?*, 62.

56. Drake and Sobel, *Is Anyone Out There?*, 52.

57. Drake and Sobel, *Is Anyone Out There?*, 62.

58. Drake and Sobel, *Is Anyone Out There?*, 64.

59. Jamieson, *Reason*, 20.

60. "The Television Infrared Observation Satellite Program (TIROS)," *NASA Science*, May 22, 2016, https://science.nasa.gov/missions/tiros/.

CHAPTER 2: WHAT THE ROBOT AMBASSADORS SAY

1. Franklin O'Donnell, "The Venus Mission: How *Mariner 2* Led the World to the Planets," Jet Propulsion Laboratory website, https://www.jpl.nasa.gov/mariner2/.

2. David R. Williams, "Chronology of Lunar and Planetary Exploration," Goddard Space Flight Center, last modified August 8, 2017, https://nssdc.gsfc.nasa.gov/planetary/chronology.html.

3. David R. Williams, "Venus Fact Sheet," Goddard Space Flight Center, last modified December 23, 2016, https://nssdc.gsfc.nasa.gov/planetary/factsheet/venusfact.html.

4. O'Donnell, "The Venus Mission."

5. Larry Klaes, "Remembering the Early Robotic Explorers," *Centauri Dreams: Imagining and Planning Interstellar Exploration*, August 29, 2012, https://www.centauri-dreams.org/?p=24285.

6. O'Donnell, "The Venus Mission."

7. O'Donnell, "The Venus Mission."

8. Williams, "Venus Fact Sheet."

9. William Sheehan and John Edward Westfall, *The Transits of Venus* (Amherst, NY: Prometheus Books, 2004), 213.

10. Sheehan and Westfall, *Transits*, 213.

11. Mikhail Ya. Marov, "Mikhail Lomonosov and the Discovery of the Atmosphere of Venus During the 1761 Transit," in *Transits of Venus: New Views of the Solar System and Galaxy, Proceedings of the 196th Colloquium of the International Astronomical Union*, ed. D.W. Kurtz (Cambridge: Cambridge University Press, 2004).

12. F. W. Taylor and D. M. Hunten, "Venus: Atmosphere," in *Encyclopedia of the Solar System*, 3rd ed., eds. Tilman Spohn, Doris Breuer, and Torrence V. Johnson (Amsterdam: Elsevier, 2014).

13. C. H. Mayer, T. P. McCullough, and R. M. Sloanaker, "Observations of Venus at 3.15 cm Wave Length," *Astrophysical Journal* 127, no. 1 (January 1958): 1–10.

14. Paolo Ulivi with David M. Harland, *Robotic Exploration of the Solar System: Part 1, The Golden Age, 1957–1982* (Berlin: Springer, 2007), xxxi.

15. Ulivi and Harland, *Robotic Exploration*, xxxii.

16. "Planetary Temperatures," Australian Space Academy, http://www.spaceacademy.net.au/library/notes/plantemp.htm.

17. Keay Davidson, *Carl Sagan: A Life* (New York: Wiley, 1999), 39–56.

18. Ray Spangenburg and Kit Moser, *Carl Sagan: A Biography* (Westport, CT: Greenwood, 2004), 11–29.

19. Kenneth R. Lang, "Global Warming: Heating by the Greenhouse Effect," *NASA's Cosmos*, 2010, http://ase.tufts.edu/cosmos/view_chapter.asp?id=21&page=1.

20. Tim Sharp, "What Is the Temperature on Earth?," *Space.com*, September 28, 2012, https://www.space.com/17816-earth-temperature.html.

21. F. W. Taylor, *Planetary Atmospheres* (Oxford: Oxford University Press, 2010), 12.

22. Svante Arrhenius, "On the Influence of Carbonic Acid in the Air upon the Temperature of the Ground," *Philosophical Magazine and Journal of Science* 41, no. 251 (April 1896): 237–76.

23. Spencer Weart, "The Carbon Dioxide Greenhouse Effect," *The Discovery of Global Warming*, January 2017, https://history.aip.org/climate/co2.htm.

24. Spangenburg and Moser, *Carl Sagan*, 36–38.

25. Davidson, *Carl Sagan*.

26. O'Donnell, "The Venus Mission."

27. Tony Reichhardt, "The First Planetary Explorers," *Air and Space Magazine*, December 14, 2012, http://www.airspacemag.com/daily-planet/the-first-planetary-explorers-162133105/.

28. O'Donnell, "The Venus Mission."

29. Asif A. Siddiqi, *Deep Space Chronicle: A Chronology of Deep Space and Planetary Probes 1958–2000* (Washington, DC: National Aeronautics and Space Administration, 2002).

30. Taylor, *Planetary Atmospheres*, 113–15.

31. Taylor, *Planetary Atmospheres*, 114–24.

32. The cold trap works by keeping water in the lower part of the atmosphere. As water vapor rises, it eventually cools and condenses, falling back to Earth. This process intensifies at the tropopause (fifteen kilometers above sea level), where air temperatures drop far below freezing. Thus, all remaining water in the atmosphere is frozen out. Michael Denton, "The Cold Trap: How It Works," *Evolution News and Science Today*, May 10, 2014, https://evolutionnews.org/2014/05/the_cold_trap_h/.

33. Davidson, *Carl Sagan*.

34. Spangenburg and Moser, *Carl Sagan*, 34–65.

35. "Mars Exploration Rovers: Step-by-Step Guide to Entry, Descent, and Landing," Jet Propulsion Laboratory, https://mars.nasa.gov/mer/mission/tl_entry1.html.

36. Steven W. Squyres, *Roving Mars: Spirit, Opportunity, and the Exploration of the Red Planet* (New York: Hyperion, 2005), 292–93.

37. Ulivi and Harland, *Robotic Exploration*, xxxiii–xxxiv.

38. Vakoch, *Astrobiology, History, and Society*, 108.

39. William Sheehan, *The Planet Mars: A History of Observation and Discovery* (Tucson, AZ: University of Arizona Press, 1996).

40. Sheehan, *Planet Mars*.

41. Rod Pyle, "Alone in the Darkness: *Mariner 4* to Mars, 50 Years Later," California Institute of Technology, July 14, 2015, https://www.caltech.edu/news/alone-darkness-mariner-4-mars-50-years-later-47324.

42. "The Dead Planet," *New York Times*, July 30, 1965.

43. Ulivi and Harland, *Robotic Exploration*, 108–12.

44. Ulivi and Harland, *Robotic Exploration*, 114–16.

45. Elizabeth Howell, "*Mariner 9*: First Spacecraft to Orbit Mars," *Space.com*, November 12, 2012, https://www.space.com/18439-mariner-9.html..

46. "Welcome to the Planets," Jet Propulsion Laboratory, https://pds.jpl.nasa.gov/planets/choices/mars1.htm.

47. Davidson, *Carl Sagan*, 279–80.

48. David R. Williams, "Viking Mission to Mars," Goddard Space Flight Center, last modified September 5, 2017, https://nssdc.gsfc.nasa.gov/planetary/viking.html, and "A Chronology of Mars Exploration," National Aeronautics and Space Administration, last modified April 16, 2015, https://history.nasa.gov/printFriendly/marschro.htm.

49. "Overview: The Mars Exploration Program," National Aeronautics and Space Administration, https://mars.nasa.gov/programmissions/overview/.

50. Robert Haberle, interview with the author, March 20, 2017.

51. "The History of Mars General Circulation Model," Mars Climate Modeling Center, https://spacescience.arc.nasa.gov/mars-climate-modeling-group/history.html.

52. Haberle, interview.

53. Williams, "Viking Mission to Mars."

54. Williams, "Viking Mission to Mars."

55. Derek Hayes, *Historical Atlas of the Pacific Northwest* (Vancouver, BC: Douglas and McIntyre, 2001).

56. Anders Persson, "Hadley's Principle: Part 1—A Brainchild with Many Fathers," *Weather* 63, no. 11 (November 2008): 335–38.

57. David R. Williams, "Mars Fact Sheet," Goddard Space Flight Center, last modified December 23, 2016, https://nssdc.gsfc.nasa.gov/planetary/factsheet/marsfact.html.

58. Haberle, interview.

59. Rob Gutro, "Polar Vortex Enters Northern U.S.," Goddard Space Flight Center, 2014, https://www.nasa.gov/content/goddard/polar-vortex-enters-northern-us/#.WcAeq62UUo-.

60. Laura Dattaro, "Check the Weather on Mars, Where NASA's MAVEN Is Headed," Weather Channel, November 19, 2013, https://weather.com/science/news/check-weather-mars-where-nasas-maven-headed-20131119.

61. Andrew P. Ingersoll, *Planetary Climates* (Princeton, NJ: Princeton University Press, 2013), 96–106.

62. National Aeronautics and Space Administration, "Minerals in Mars 'Berries' Adds to Water Story," news release, March 18, 2004, https://mars.nasa.gov/mer/newsroom/pressreleases/20040318a.html.

63. National Aeronautics and Space Administration, "NASA Rover Finds Old Streambed on Martian Surface," news release, September 27, 2012, https://www.nasa.gov/mission_pages/msl/news/msl20120927.html.

64. Michael H. Carr and James W. Head III, "Geologic History of Mars," *Earth and Planetary Science Letters* 294, nos. 3–4 (June 1, 2010): 185–203.

65. Paul L. Montgomery, "Throngs Fill Manhattan to Protest Nuclear Weapons," *New York Times*, June 13, 1982.

66. Robert S. Norris and Hans M. Kristensen, "Global Nuclear Weapons Inventories, 1945–2010," *Bulletin of the Atomic Scientists* 66, no. 4 (July/August 2010): 77–83.

67. R. P. Turco et al., "Nuclear Winter: Global Consequences of Multiple Nuclear Explosions," *Science* 222, no. 4630 (December 23, 1983): 1283–92.

68. Jill Lepore, "The Atomic Origins of Climate Science," *The New Yorker*, January 30, 2017, http://www.newyorker.com/magazine/2017/01/30/the-atomic-origins-of-climate-science.

69. Jacob Darwin Hamblin, "Badash, *A Nuclear Winter's Tale*," *Metascience* 21, no. 3 (November 2012): 727–31.

CHAPTER 3: THE MASKS OF EARTH

1. "Earth's Early Atmosphere," *Astrobiology Magazine*, December 2, 2011, http://www.astrobio.net/geology/earths-early-atmosphere/.

2. John Reed, "Inside the Army's Secret Cold War Ice Base," Defense Tech, April 6, 2012, https://www.defensetech.org/2012/04/06/inside-the-armys-secret-cold-war-ice-base/, and Malcolm Mellor, *Oversnow Transport* (Hanover, NH: U.S. Army Cold Regions Research and Engineering Laboratory, 1963), http://www.dtic.mil/dtic/tr/fulltext/u2/404778.pdf.

3. "Trail Blazed by Renowned Explorer Leads Danish, U.S. Scouts to Arctic Adventure," *Army Research and Development*, December 1960, 14.

4. "The Ice Sheet," *Visit Greenland*, http://www.greenland.com/en/about-greenland/nature-climate/the-ice-cap/.

5. Frank J. Leskovitz, "Camp Century, Greenland: Science Leads the Way," http://gombessa.tripod.com/scienceleadstheway/id9.html.

6. Leskovitz, "Camp Century."

7. Leskovitz, "Camp Century."

8. Leon E. McKinney, "Camp Century Greenland," West-Point.org, http://www.west-point.org/class/usma1955/D/Hist/Century.htm.

9. Gordon de Q. Robin, "Profile Data, Greenland Region," in *The Climate Record in Polar Ice Caps*, ed. Gordon de Q. Robin (1983; repr., Cambridge: Cambridge University Press, 2010), 100–101.

10. Joseph Gale, *Astrobiology of Earth: The Emergence, Evolution, and Future of Life on a Planet in Turmoil* (Oxford: Oxford University Press, 2009), 125–26.

11. John S. Schlee, "Our Changing Continent," United States Geological Survey, last modified February 15, 2000, https://pubs.usgs.gov/gip/continents/.

12. Gale, *Astrobiology of Earth*, 125.

13. Willi Dansgaard, *Frozen Annals: Greenland Ice Cap Research* (Copenhagen: Niels Bohr Institute, 2005), 55–56.

14. Dansgaard, *Frozen Annals*, 58.

15. W. Dansgaard et al., "One Thousand Centuries of Climate Record from Camp Century on the Greenland Ice Sheet," *Science* 166, no. 3903 (October 17, 1969): 377–80. See also "The Younger Dryas," NOAA National Centers for Environmental Information, https://www.ncdc.noaa.gov/abrupt-climate-change/The%20Younger%20 Dryas.

16. Manned Spacecraft Center, "*Apollo 8* Onboard Voice Transcription, As Recorded on the Spacecraft Onboard Recorder (Data Storage Equipment)," January 1969, 113–14, https://www.jsc.nasa.gov/history/ mission_trans/AS08_CM.PDF.

17. *Earthrise, Time,* http://100photos.time.com/photos/nasa-earthrise -apollo-8.

18. K. M. Cohen, S. Finney, and P. L. Gibbard, "International Chronostratigraphic Chart," *International Commission on Stratigraphy*, January 2013, http://www.stratigraphy.org/icschart/chronostratchart 2013-01.pdf.

19. Ann Zabludoff, "Lecture 13: The Nebular Theory of the Origin of the Solar System," University of Arizona Department of Astronomy and Steward Observatory, http://atropos.as.arizona.edu/aiz/teaching/ nats102/mario/solar_system.html.

20. C. Goldblatt et al., "The Eons of Chaos and Hades," *Solid Earth Discussions* 1, no. 1 (2010), 1–3.

21. Goldblatt et al., "Chaos and Hades."

22. Thomas Holtz, "GEOL 102 Historical Geology: The Archean Eon," University of Maryland Department of Geology, last modified January 18, 2017, https://www.geol.umd.edu/~tholtz/G102/lectures/ 102archean.html.

23. Stanly M. Awramik and Kenneth J. McNamara, "The Evolution and Diversification of Life," in *Planets and Life: The Emerging Science of Astrobiology*, eds. Woodruff R. Sullivan III and John A Baross (Cambridge University Press, 2007), 313–16.

24. Awramik and McNamara, "Evolution and Diversification" 313–18.

25. Z. X. Li et al., "Assembly, Configuration, and Break-up History of Rodinia: A Synthesis," *Precambrian Research*, 160 (2008): 179–210.

26. David Catling and James F. Kasting, "Planetary Atmospheres and Life," in *Planets and Life: The Emerging Science of Astrobiology*, eds. Woodruff R. Sullivan III and John A Baross (Cambridge University Press, 2007), 99.

27. Awramik and McNamara, "Evolution and Diversification," 321.

28. Donald E. Canfield, *Oxygen: A Four Billion Year History* (Princeton, NJ: Princeton University Press, 2014), 145–46.

29. "PETM: Global Warming, Natural," *Weather Underground*, https://www.wunderground.com/climate/PETM.asp?MR=1.

30. Canfield, *Oxygen*, 13.

31. Canfield, *Oxygen*, 14.

32. "Opening a Tectonic Zipper," *Seismo Blog* (UC Berkeley Seismology Lab), April 5, 2010, http://seismo.berkeley.edu/blog/2010/04/05/opening-a-tectonic-zipper.html.

33. Canfield, *Oxygen*, 14.

34. Canfield, *Oxygen*, 14.

35. Canfield, *Oxygen*, 41.

36. Canfield, *Oxygen*, 41.

37. Gale, *Astrobiology of Earth*, 110–11.

38. Canfield, *Oxygen*, 41–42.

39. David C. Catling, *Astrobiology: A Very Short Introduction* (Oxford: Oxford University Press, 2013), 50–55.

40. Catling, *Astrobiology*, 52.

41. Alexej M. Ghilarov, "Vernadsky's Biosphere Concept: An Historical Perspective," *Quarterly Review of Biology* 70, no. 2 (June 1995): 193–203.

42. Irina Trubetskova, "Vladimir Ivanovich Vernadsky and His Revolutionary Theory of the Biosphere and the Noosphere," University of New Hampshire, http://www-ssg.sr.unh.edu/preceptorial/Summaries_2004/Vernadsky_Pap_ITru.html.

43. Ghilarov, "Vernadsky's Biosphere Concept."

44. Vladimir Vernadsky, *The Biosphere*, trans. David B. Langmuir (New York: Copernicus, 1998), 44, 56.

45. Ghilarov, "Vernadsky's Biosphere Concept."

46. James Lovelock, *Homage to Gaia: The Life of an Independent Scientist* (New York: Oxford University Press, 2000).

47. Lovelock, *Homage to Gaia*, 242.

48. Lovelock, *Homage to Gaia*, 243.

49. Lovelock, *Homage to Gaia*, 243.

50. Lovelock, *Homage to Gaia*, 243–44.

51. Lovelock, *Homage to Gaia*, 253.

52. Lovelock, *Homage to Gaia*, 255.

53. Joel Bartholomew Hagen, Douglas Allchin, and Fred Singer, *Doing Biology* (New York: HarperCollins, 1996).

54. Lovelock, *Homage to Gaia*, 256–57.

55. Michael Ruse, "Earth's Holy Fool?," *Aeon*, https://aeon.co/essays/gaia-why-some-scientists-think-it-s-a-nonsensical-fantasy.

56. John Postgate, "Gaia Gets Too Big for Her Boots," *New Scientist*, April 7, 1988.

57. Ruse, "Earth's Holy Fool?"

58. Lovelock, *Homage to Gaia*, 265.

59. Ruse, "Earth's Holy Fool?"

CHAPTER 4: WORLDS BEYOND MEASURE

1. This section on Thomas See is based on information found in Thomas J. Sherrill, "A Career of Controversy: The Anomaly of T.J.J. See," *Journal for the History of Astronomy* 30, no. 1 (February 1999): 25–50, and William Sheehan, "The Tragic Case of T.J.J. See," *Mercury* 31, no. 6 (November 2002): 34.

2. Personal correspondence.

3. Amy Veltman, "Dr. Jill Tarter: Looking to Make 'Contact,'" *Space.com*, November 12, 1999, https://web.archive.org/web/20081005020231/http://www.space.com/peopleinterviews/tarter_profile_991112.html.

4. "Jill Tarter," SETI Institute, https://www.seti.org/users/jill-tarter.

5. Jill Tarter, interview with the author.

6. John Billingham, "SETI: The NASA Years," in *Searching for Extrater-*

restrial Intelligence: SETI Past, Present, and Future, ed. H. Paul Shuch (Berlin: Springer, 2011), 70.

7. Jesse L. Greenstein and David C. Black, "Detection of Other Planetary Systems," in *The Search for Extraterrestrial Intelligence: SETI*, eds. Philip Morrison, John Billingham, and John Wolfe (Washington, DC: NASA Scientific and Technical Information Office, 1977).

8. Greenstein and Black, "Detection."

9. Tarter, interview.

10. David C. Black and William E. Brunk, eds., *An Assessment of Ground-Based Techniques for Detecting Other Planetary Systems, Volume 1: An Overview* (Moffett Field, CA: National Aeronautics and Space Administration, 1979), 18.

11. Michael D. Lemonick, *Mirror Earth: The Search for Our Planet's Twin* (New York: Walker, 2012), 55.

12. Lemonick, *Mirror Earth*.

13. Lemonick, *Mirror Earth*, 52–53.

14. Lemonick, *Mirror Earth*, 58.

15. Andrew Lawler, "Bill Borucki's Planet Search," *Air and Space*, May 2003, http://www.airspacemag.com/space/bill-boruckis-planet-search -4545405/?no-ist.

16. Lawler, "Borucki's Planet Search."

17. Lawler, "Borucki's Planet Search."

18. William J. Borucki et al., "Kepler Planet-Detection Mission: Introduction and First Results," *Science* 327, no. 5968 (February 19, 2010): 977–80.

19. "Liftoff of Kepler: On a Search for Exoplanets in Some Way Like Our Own," National Aeronautics and Space Administration, March 6, 2009, https://www.nasa.gov/multimedia/imagegallery/image_ feature_2123.html.

20. Natalie Batalha, interview with the author.

21. Michele Johnson, "NASA's Kepler Mission Announces a Planet Bonanza, 715 New Worlds," National Aeronautics and Space Administration, February 26, 2014, https://www.nasa.gov/ames/kepler/nasas -kepler-mission-announces-a-planet-bonanza.

22. "Exoplanet Anniversary: From Zero to Thousands in 20 Years," Jet

Propulsion Laboratory, October 6, 2015, https://www.jpl.nasa.gov/news/news.php?feature=4733.

23. Batalha, interview.

24. "Star: KOI-961—3 PLANETS," *Extrasolar Planets Encyclopaedia*, http://exoplanet.eu/catalog/?f='KOI-961'+in+name.

25. Lee Billings, "Newfound Super-Earth Boosts Search for Alien Life," *Scientific American*, April 19, 2017, https://www.scientificamerican.com/article/newfound-super-earth-boosts-search-for-alien-life/.

26 Shannon Hall, "This Super-Saturn Alien Planet Might Be the New 'Lord of the Rings,'" *Space.com*, February 3, 2015, https://www.space.com/28435-super-saturn-alien-planet-rings.html.

27. Andrew Fazekas, "Diamond Planet Found—Part of 'Whole New Class'?," *National Geographic*, October 13, 2012, http://news.nationalgeographic.com/news/2012/10/121011-diamond-planet-space-solar-system-astronomy-science/.

28. "Hubble Finds a Star Eating a Planet," Hubble Space Telescope, May 20, 2010, https://www.nasa.gov/mission_pages/hubble/science/planet-eater.html.

29. Amelie Saintonge, "How Many Stars Are Born and Die Each Day?," *Ask An Astronomer*, last modified June 27, 2015, http://curious.astro.cornell.edu/about-us/83-the-universe/stars-and-star-clusters/star-formation-and-molecular-clouds/400-how-many-stars-are-born-and-die-each-day-beginner.

30. Mike Wall, "Nearly Every Star Hosts at Least One Alien Planet," *Space.com*, March 4, 2014, https://www.space.com/24894-exoplanets-habitable-zone-red-dwarfs.html.

31. Robert Sanders, "Astronomers Answer Key Question: How Common Are Habitable Planets?" University of California, Berkeley, November 4, 2013, http://news.berkeley.edu/2013/11/04/astronomers-answer-key-question-how-common-are-habitable-planets/.

32. In our paper, we set the pessimism line between 10^{-24} and 10^{-22} for technical reasons. Throughout the book, we will use the more conservative value of 10^{-22}.

33. Ross Andersen, "Fancy Math Can't Make Aliens Real," *Atlantic*, June 17, 2016, https://www.theatlantic.com/science/archive/2016/06/

fancy-math-cant-make-aliens-real/487589/, and Ethan Siegel, "Humanity May Be Alone in the Universe," *Forbes*, June 21, 2016, https://www.forbes.com/sites/startswithabang/2016/06/21/humanity-may-be-alone-in-the-universe/.

34. Adam Frank, "Yes, There Have Been Aliens," *New York Times*, June 10, 2016, https://www.nytimes.com/2016/06/12/opinion/sunday/yes-there-have-been-aliens.html.

35. Ernst Mayr, "Can SETI Succeed? Not Likely," The Planetary Society, http://daisy.astro.umass.edu/~mhanner/Lecture_Notes/Sagan-Mayr.pdf.

36. Brandon Carter, "The Anthropic Principle and its Implications for Biological Evolution," *Philosophical Transactions of the Royal Society A* 310, no. 1512 (December 1983): 347–63.

37. It's worth noting that authors like astrophysicist Mario Livio have presented arguments that undermine the basis for Carter's work. Mario Livio, "How Rare Are Extraterrestrial Civilizations, and When Did They Emerge?" *The Astrophysical Journal* 511, no. 1 (1999): 429–31.

38. Hubert P. Yockey, "A Calculation of the Probability of Spontaneous Biogenesis by Information Theory," *Journal of Theoretical Biology* 67, no. 3 (August 7, 1977): 377–98.

39. Wentao Ma et al., "The Emergence of Ribozymes Synthesizing Membrane Components in RNA-Based Protocells," *Biosystems* 99, no. 3 (March 2010): 201–9.

CHAPTER 5: THE FINAL FACTOR

1. William Bains. "Many Chemistries Could Be Used to Build Living Systems," *Astrobiology* 4, no. 2 (June 2004): 137–67.

2. J. R. Haas, "The Potential Feasibility of Chlorinic Photosynthesis on Exoplanets," *Astrobiology* 10, no. 9 (November 2010): 953–63.

3. J. Dulcic, A. Soldo, and I. Jardas, "Adriatic Fish Biodiversity and Review of Bibliography Related to Croatian Small-Scale Coastal Fisheries," http://www.faoadriamed.org/pdf/publications/td15/wp_dulcica.pdf.

4. Sharon Kingsland, *Modeling Nature: Episodes in the History of Population Ecology* (Chicago: University of Chicago Press, 1985), 106.

5. Philip J. Davis, "Carissimo Papà: A Great Fish Story," *SIAM News* 38, no. 8 (October 2005).

6. Kingsland, *Modeling Nature*, 4.

7. Mark Kot, *Elements of Mathematical Ecology* (Cambridge: Cambridge University Press, 2001), 11.

8. Kingsland, *Modeling Nature*, 109.

9. Kingsland, *Modeling Nature*, 106–15.

10. Kingsland, *Modeling Nature*, 1.

11. Rafael Reuveny, "Taking Stock of Malthus: Modeling the Collapse of Historical Civilizations," *Annual Review of Resource Economics* 4 (2012): 303–29.

12. Reuveny, "Taking Stock of Malthus," 303.

13. Erich von Däniken, *Chariots of the Gods?* (1968; New York: Putnam, 1970).

14. Jared Diamond, *Collapse: How Societies Choose to Fail or Succeed* (New York: Viking, 2005).

15. James A. Brander and M. Scott Taylor, "The Simple Economics of Easter Island: A Ricardo-Malthus Model of Renewable Resource Use," *American Economic Review* 88, no. 1 (March 1998): 119–38.

16. Bill Basener and David S. Ross, "Booming and Crashing Populations and Easter Island," *SIAM Journal on Applied Mathematics* 65, no. 2 (2004): 684–701.

17. Adam Frank and Woodruff Sullivan, "Sustainability and the astrobiological perspective," *Anthropocene* 5 (March 2014): 32.

18. Adam Frank, "Could You Power Your Home With A Bike?," NPR, December 8, 2016, http://www.npr.org/sections/13.7/2016/12/08/50479 0589/could-you-power-your-home-with-a-bike.

19. Rudy M. Baum Sr., "Future Calculations: The First Climate Change Believer," *Distillations*, Summer 2016, https://www.chemheritage.org/ distillations/magazine/future-calculations.

20. L. Miller, F. Gans, and A. Kleidon, "Estimating Maximum Global Land Surface Wind Power Extractability and Associate Climatic Consequences," *Earth System Dynamics* 2 (2011): 112.

21. It's worth mentioning that it is possible that the distribution of exocivilizations might be more complicated than providing a well-defined

average. There might, for example, be two peaks in the lifetimes of a large sample of exo-civilizations (one short and long). This kind of result would be interesting in its own right.

CHAPTER 6: THE AWAKENED WORLDS

1. Marina Alberti, *Cities That Think Like Planets* (Seattle: University of Washington Press, 2016).

2. Drake and Sobel, *Is Anyone Out There?*.

3. Drake and Sobel, *Is Anyone Out There?*. Also, "First Soviet-American Conference on Communication with Extraterrestrial Intelligence," *Icarus* 16, no. 2 (April 1972): 412.

4. Kenneth I. Kellermann, "Nicolay Kardashev," National Radio Astronomy Observatory, http://rahist.nrao.edu/kardashev_reber-medal.shtml.

5. Nikolai Kardashev, "Transmission of Information by Extraterrestrial Civilizations," *Soviet Astronomy* 8, no. 2 (September/October 1964): 217, and Milan M. Cirkovic, "Kardashev's Classification at 50+: A Fine Vehicle with Room for Improvement," *Serbian Astronomical Journal* 191 (2015): 1–15.

6. "Energy of a Nuclear Explosion," *The Physics Factbook*, https://hypertextbook.com/facts/2000/MuhammadKaleem.shtml.

7. Freeman J. Dyson, "Search for Artificial Stellar Sources of Infrared Radiation," *Science* 131, no. 3414 (June 3, 1960): 1667–68.

8. J. T. Wright et al., "The Ĝ Infrared Search for Extraterrestrial Civilizations with Large Energy Supplies, II. Framework, Strategy, and First Result," *Astrophysical Journal* 792, no. 1 (2014): 27.

9. Cirkovic, "Kardashev's Classification."

10. Carl Sagan, *Carl Sagan's Cosmic Connection: An Extraterrestrial Perspective*, ed. Jerome Agel (Cambridge: Cambridge University Press, 2000).

11. Michio Kaku, "The Physics of Extraterrestrial Civilizations," http://mkaku.org/home/articles/the-physics-of-extraterrestrial-civilizations/.

12. Isaac Asimov, *Foundation* (New York: Gnome Press, 1951).

13. Second Law of Thermodynamics, http://hyperphysics.phy-astr.gsu.edu/hbase/thermo/seclaw.html.

14. Matt Williams, "What is the Weather Like on Mercury?," *Universe Today*, July 24, 2017, https://www.universetoday.com/85348/weather-on-mercury/.

15. Volatility is a concept from physics and chemistry and is the tendency of a substance to vaporize. Volatiles in planetary science are substances that will vaporize (or boil) at "normal" temperatures and pressures. Iron, for example, is not considered a volatile, while water, CO_2, and methane are volatiles.

16. L. Kaltenegger and D. Sasselov, "Detecting Planetary Geochemical Cycles on Exoplanets: Atmospheric Signatures and the Case of SO_2," *Astrophysical Journal* 708, no. 2 (2010): 1162–67, and J. F. Kasting and D. E. Canfield, "The Global Oxygen Cycle," in *Fundamentals of Geobiology*, eds. A. H. Knoll, D. E. Canfield, and K. O. Konhauser (Hoboken, NJ: Wiley-Blackwell, 2012), 93–104.

17. Adam Frank, Axel Kleidon, and Marina Alberti, "Earth as a Hybrid Planet: The Anthropocene in an Evolutionary Astrobiological Context," *Anthropocene* (forthcoming).

18. Donald Canfield, "The Early History of Atmospheric Oxygen," *Annual Review of Earth and Planetary Sciences* 33 (2005): 1–36.

19. Eleni Stavrinidou et al., "Electronic Plants," *Science Advances* 1, no. 10 (November 2015).

20. David Grinspoon, *The Earth in Human Hands* (New York: Grand Central Publishing, 2016).

21. Based on Vernadsky's writings, the Jesuit priest and paleontologist Pierre Teilhard de Chardin worked on his own, decidedly more mystical version of the noosphere. P. Teilhard de Chardin, *The Phenomenon of Man*, trans. Bernard Wall (New York: Harper, 1959), 238.

22. The "Big History Project" is an attempt to teach history in a way that puts humanity in its place along with the rest of the cosmos. See https://www.bighistoryproject.com/home.

ILLUSTRATION CREDITS

INDEX

Italic page numbers refer to illustrations.